Nimble with Numbers

Engaging Math Experiences to Enhance Number Sense and Promote Practice

Grades 6 and 7

Leigh Childs, Laura Choate, and Polly Hill

Dale Seymour Publications®
White Plains, New York

Acknowledgments

Thanks to special friends and colleagues who helped field test these activities with their students and who gave valuable feedback and suggestions:

Maryann Wickett

Eunice Hendrix-Martin

Polly Hill

Kathy Grulich

Thanks to super kids who provided a student perspective as they tried these activities:

Nathan Burton

Ryan Choate

Dan Hill

Managing Editor: Catherine Anderson
Senior Editors: Carol Zacny, John Nelson
Project Editor: Patsy Norvell
Consulting Editor: Nancy Anderson
Production/Manufacturing Director: Janet Yearian
Production/Manufacturing Manager: Karen Edmonds
Production/Manufacturing Coordinator: Joan Lee
Design Manager: Jeff Kelly
Text design: Tani Hasegawa
Page composition: Susan Cronin-Paris
Cover design and illustration: Ray Godfrey
Art: Stefani Sadler

This book is published by Dale Seymour Publications®, an imprint of Addison Wesley Longman, Inc.

Dale Seymour Publications
10 Bank Street
White Plains, NY 10602
Customer Service: 800-872-1100

Order number DS21863
ISBN 1-57232-986-6

1 2 3 4 5 6 7 8 9 10-VG-02-01-00-99-98

Table of Contents

Integers

Algebra

Blackline Masters

Introduction to *Nimble with Numbers*

Why This Book?

National recommendations for a meaning-centered, problem-solving mathematics curriculum place new demands on teachers, students, and parents. Students need a facility with numbers and operations to achieve success in today's mathematics programs. Basics for students today require a broadening of the curriculum to include all areas of mathematics. Students are being asked to demonstrate proficiency not just in skills, but in problem solving, critical thinking, conceptual understanding, and performance tasks. Consequently, the reduced time teachers devote to number must be thoughtful, selective, and efficient.

This book fulfills the need for high-quality, engaging math experiences that provide meaningful practice and further the development of number and operation sense. **These activities are designed to help students practice number and pre-algebra concepts previously taught for understanding in a variety of contexts.** Besides meeting the need for effective practice, *Nimble with Numbers*

- Provides a variety of adaptable formats for essential practice
- Supplements and enhances homework assignments
- Encourages parents' involvement in developing their students' proficiency with mental computation, fractions, decimals, percents, integers, and algebraic expressions
- Provides motivating and meaningful lessons for a substitute teacher

Criteria for Preferred Activities

For efficient use of time devoted to number-related topics, the book focuses on activities that are

- Inviting (encourages participation)
- Engaging (maintains interest)
- Simple to learn
- Repeatable (able to reuse often and sustain interest)
- Open-ended, allowing multiple solutions
- Easy to prepare
- Easy to adapt for various levels
- Easy to vary for extended use

In addition, these activities

- Require a problem-solving approach
- Improve basic skills
- Enhance number sense and operation sense
- Encourage strategic thinking
- Promote mathematical communication
- Promote positive attitudes toward mathematics as abilities improve

Planning Made Simple

Organization of Book

The activities of this book are divided into seven sections that cover high-priority, number-related topics for sixth graders and seventh graders. The first section promotes mental computation of all operations. The second section reviews number theory topics such as primes, common factors, and least common multiples.

The third, fourth, and fifth sections reinforce fraction, decimal, and percent concepts with an effort to relate these three topics, especially in the Percent section. When students seem confident with the first five sections, proceed to the more challenging sixth and seventh sections. The Integers section emphasizes operations, and the Algebra section reinforces modeling and solving simpler algebraic equations.

Each section begins with an overview and suggestions to highlight the activities and provide some time-saving advice. The interactive activities identify the specific topic practiced (Topic), the objective (Object), the preferred grouping of participants (Groups), and the materials required (Materials). Activities conclude with "Making Connections" questions to promote reflection and help students make mathematical connections. "Tips" provide helpful implementation suggestions and variations. Needed blackline masters are included with the activity or in the Blackline Masters section at the end of the book.

This introductory section includes a matrix of activities. The repeatable Sponges and Games are listed alphabetically with corresponding information to facilitate their use.

Types of Activities

The book contains activities for whole group, small groups, pairs, and individuals. Each section provides

- Sponges (S)
- Skill Checks (C)
- Games (G)
- Independent Activities (I)

Sponges

Sponges are enriching activities for soaking up spare moments. Use Sponges as warm-up or spare-time activities with the whole class or with small groups. Sponges usually require little or no preparation and are short in duration (3–15 minutes). These appealing Sponges are repeatable and, once they become familiar, can be student-led.

Skill Checks

The Skill Checks in each section provide a way to show students' improvement to parents as well as to themselves. Each page is designed to be duplicated and cut in half, providing six comparative records for each student. Before solving the ten problems, students should respond to the starter task following the STOP sign. These starter tasks are intended to promote mental computation and build number sense. Some teachers believe their students perform better on the Skill Checks if the responses to the STOP task are shared and discussed before students solve the remaining problems. Most students will complete a Skill Check in 10 to 15 minutes. The concluding extension problem, labeled "GO ON," accommodates those students who finish early. We recommend that early finishers be encouraged to create similar problems for others to solve. By having students discuss their approaches and responses, teachers help students discover more efficient mental computation strategies.

Games

Initially a new Game might be modeled with the entire class, even though Games are intended to be played by pairs or small groups after the rules are understood. An excellent option is to share the Game with a few students who then teach the Game to others. To facilitate getting started, teachers may recommend some procedure for identifying the first player or pair. A recommended arrangement is to have players partner up and collaborate as "pair players" against another pair. Mathematical thinking and communication are enhanced as students collaborate to develop and share successful strategies. Many Games provide easier and more challenging versions. Most Games require approximately 20 to 45 minutes of playing time. Games are ideal for home use since they provide students with additional practice and reassure parents that the "basics" continue to be valued. When sending gameboards home, be sure to include the directions.

Games and Sponges provide students with a powerful vehicle for assessing their own mathematical abilities. During the Games, students receive immediate feedback that allows them to revise and to correct inefficient and inadequate practices. Sponges and Games differ from the Independent Activities since they usually need to be introduced by a leader.

Independent Activities

Independent Activity sheets provide computation practice and review of number-related topics and basic algebraic concepts. These sheets are designed to encourage practice of many more number concepts than would seem apparent at first glance. Some Independent Activities allow multiple solutions. Most students will complete an Independent Activity sheet in 15 to 30 minutes. Independent Activity sheets can be completed in class or sent home as homework. There are two versions of several Independent Activities to accommodate different levels of difficulty. In the Fractions, Decimals, and Integers sections, the two versions of some Independent Activities allow practice of addition and subtraction or multiplication and division. Many Independent Activities can be easily modified to provide additional practice.

Suggestions for Using *Nimble with Numbers*

Materials Tips

An effort has been made to minimize the materials needed. When appropriate, blackline masters are provided. The last section of the book contains more generic types of blackline masters. The six-sectioned spinners (p. 150) can substitute for a number cube or die. The blank spinner can be used for the specially marked number cubes (3–8, 4–9 or 3, 3, 4, 4, 5, 5). A simple spinner, like the one shown, can be assembled using one of the blackline master spinner bases, a paper clip, and a pencil.

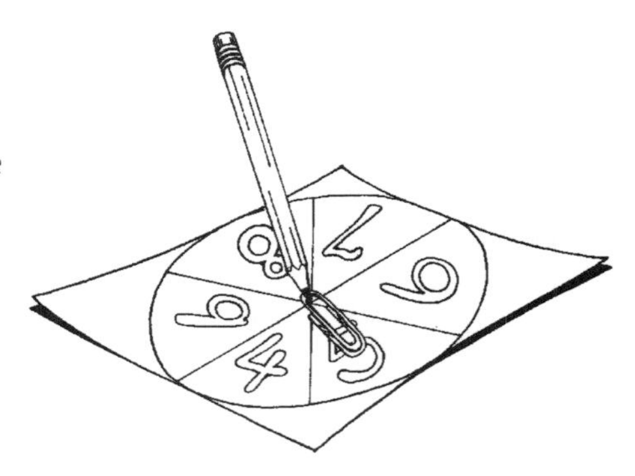

Many activities use the *Digit Squares* (p. 147). The familiar sets of 0–9 number tiles substitute well for Digit Squares. If not available, take time now to duplicate a Digit Square set on card stock for each student. By removing the zero Digit Square, students have a set of Positive Integer Squares to accompany the *Negative Integer Squares* (p. 148).

A few activities and class Sponges require *Digit Cards* (p. 146). Digit Card sets should also be duplicated on card stock. Teachers should cut two sets of Digit Cards apart, place them in an appropriate container (paper sack, coffee can, or margarine tub), and store them in a handy place.

Various materials work as markers on gameboards—different types of beans, multicolored cubes, buttons, counters, and transparent bingo chips (our preference due to the see-through feature). For some activities, students will need scratch paper and pencils. It is assumed that an overhead projector is available, but a chalkboard or whiteboard may be substituted.

Recommended Uses

The repeatable nature of the activities makes them ideal for continued use at home. Encouraging students to use these activities at home serves a dual purpose: parents are able to assist their children in gaining competence with the facts and with mental computation, and parents are reassured as they see the familiar basics practiced. To support your work in this area, we have included a parent letter and a list of helpful open-ended questions.

Besides being a source for more familiar homework, the activities offer a wide variety of classroom uses. The activities can be effectively used by substitute teachers, as rainy-day options, or for a change of pace. Many activities are short-term and require little or no preparation, making them ideal for soaking up spare moments at the end or beginning of a class period. They also work well as choices for *center* or *menu* activities. When students are absent from school, you may include these activities in independent work packets. Package these activities in manila envelopes or self-closing transparent bags to facilitate frequent and easy checkout. To modify the activities and to accommodate the needs of your students, you may easily change the numbers, operations, and directions.

Getting the Most from These Activities

It is important to focus on increasing students' awareness of the mathematics being learned. To do this, pose open-ended questions that promote reflection, communication, and mathematical connections. To assist you with this vital task, "Making Connections" questions are included with each Sponge and Game.

Having students work together as pair players is of great value in increasing student confidence. While working this way, students have more opportunities to communicate strategies and to verbalize thinking. When asked to identify and to share their successful Game strategies verbally and in writing, students grow mathematically. Also, it is worthwhile to ask students to improve these activities or to create different versions of high-interest Games.

Good questions help students make sense of mathematics, build their confidence, and encourage mathematical thinking and communication. A list of helpful sample questions appears on page 8. Since the teacher's or parent's response impacts learning, we have included suggestions for responding. Share this list with parents for their use as they assist students with these activities and other unfamiliar homework tasks. This list was created by Leigh Childs for parent workshops and for inclusion in the California Mathematics Council's *They're Counting on Us, A Parent's Guide to Mathematics Education.* We have adapted the list for use with this book.

Concluding Thought

We hope that by using these materials, your students will develop more positive feelings towards mathematics as they improve their confidence and number competence.

Parent Support

Since most parents place a high priority on attention to number-related topics, they will appreciate the inviting and repeatable activities in this book. Because most parents are willing to share the responsibility for repeated, short periods of practice, the *Family Letter* (p. 7) and *Questions Sampler* (p. 8) are designed to promote parent involvement. The first home packet might include the *Family Letter,* the *Questions Sampler,* and *Products Bingo* (pp. 21–23). Since many students might benefit by reviewing the basic facts, *Compute and Capture* (p. 19) might be considered for a home packet. This packet should include materials for making three sets of Digit Cards (p. 146). Advise students and their families to keep the Digit Cards in a safe place for repeated use during the school year.

Students enjoy and benefit from repeated use of *Compute and Score* (pp. 12–13) and *Fitting Fractions* (pp. 48–49). These Sponges lend themselves to home packets as well. Often, popular Sponges can easily be converted into Games. Students can be challenged to assist with their creation. The advantage of Sponges is that unlike Games, many of them can be experienced while a monitoring family member prepares dinner, packs lunches, or attends to other household tasks.

Family Letter

Dear Family,

To be prepared to work in the 21st century, all students need to be confident and competent in mathematics. Today the working world requires understanding of all areas of mathematics including statistics, logic, geometry, and probability. To be successful in these areas, students must maintain their basic facts and be able to compute. It is important that we be more efficient and effective in the time we devote to number-related topics. You can help your child in this area.

Throughout the school year, our mathematics program will focus on enhancing your child's understanding of number concepts. However, children must devote time at school and at home to practice and to improve these skills. Periodically, I will send home activities and related activity sheets that will build number sense and provide much needed practice. These games and activities have been carefully selected to engage your child in practicing more math skills than are usually answered on a typical page of drill.

By using the enclosed *Questions Sampler* during homework sessions, you will be able to assist your child without revealing the answers. The questions are categorized to help you select the most appropriate questions for your situation. If your child is having difficulty getting started with a homework assignment, try one of the questions in the first section. If your child gets stuck while completing a task, ask one of the questions from the second section. Try asking one of the questions from the third section to have your child clarify his or her mathematical thinking or to reflect on reasonableness of the results.

Good questions will help your child make sense of the mathematics, build confidence, and improve mathematical thinking and communication. I recommend posting the *Questions Sampler* in a convenient place so that you can refer to it often while helping your child with homework.

Your participation in this crucial area is most welcome.

Sincerely,

Questions Sampler

Getting Started

How might you begin?

What do you know now?

What do you need to find out?

While Working

How can you organize your information?

How can you make a drawing (model) to explain your thinking?

What approach (strategy) are you developing to solve this?

What other possibilities are there?

What would happen if . . . ?

What do you need to do next?

What assumptions are you making?

What patterns do you see? . . . What relationships?

What prediction can you make?

Why did you . . . ?

Checking Your Solutions

How did you arrive at your answer?

Why do you think your solution is reasonable?

What did you try that didn't work?

How can you convince me your solution makes sense?

Expanding the Response

(To help clarify your child's thinking, avoid stopping when you hear the "right" answer and avoid correcting the "wrong" answer. Instead, respond with one of the following.)

Why do you think that?

Tell me more.

In what other way might you do that? What other possibilities are there?

How can you convince me?

Matrix of Games and Sponges

Type	Title	Topic	Page	Materials	Class	Groups	Pairs
S	**Common Factors Match**	Common Factors	32	Overhead pens, Form p. 33	✓	✓	
G	**Compute and Capture**	Mental Facts and Operations	19	Digit Cards, Paper strips			✓
S	**Compute and Score**	Mental Facts and Operations	12	Digit Squares including Transparent Set, Form p. 13	✓	✓	
G	**Cover It!**	Operations with Integers	117	Number Cubes, Markers, Gameboard p. 118			✓
G	**Create and Solve**	Writing and Solving Equations	138	Number Cubes, Markers, Gameboard p. 139			✓
S	**Creating Equations**	Creating Equations for Unknowns	132	Number Cubes	✓	✓	
S	**Decimal Draw**	Adding and Subtracting Decimals	66	Transparent Digit Squares, Form p. 67	✓	✓	
G	**Decimal Four-in-a-Row**	Mentally Multiplying Decimals	75	Markers, Paper Clips, Gameboard p. 76			✓
S	**Decimal Tic-Tac-Toe**	Converting Fractions to Decimals	70	Markers, Transparent Digit Squares, Form p. 71	✓	✓	
G	**Decimal Bingo**	Mental Division, Decimal Quotients	77	Markers, Calculator, Gameboards pp. 78–79			✓
G	**Draw 4**	Mental Multiplication	20	Digit Squares			✓
S	**Exploring Percent Relationships**	Percents and Related Amounts	90	Percent Problems p. 91	✓	✓	
G	**Factor Find**	Factors and Multiples	41	Markers, Gameboard p. 42			✓
G	**Finding Unknowns**	Solving an Equation	136	Transparent Chips, Markers, Gameboard p. 137			✓
S	**Fitting Fractions**	Computing with Fractions	48	Digit Squares including Transparent Set, Form p. 49	✓	✓	
G	**Fraction Formulation**	Mental Computing with Fractions	56	Digit Squares, Forms pp. 57–58			✓
S	**Fraction Three-in-a-Row**	Fraction Sense and Computation	50	Markers, Form p. 51, Transparent Clues, p. 52	✓	✓	
S	**Fractions to Percents**	Converting Fractions to Percents	87	Digit Squares including Transparent Set, Form p. 88	✓	✓	
S	**Greatest Common Factor Bingo**	Greatest Common Factor	28	Markers, Paper Clips and Pencil, Form p. 29	✓	✓	
S	**How?**	Operations with Integers	110	5 × 8 cards, Form p. 111	✓	✓	
G	**Integer Four-in-a-Row**	Subtracting Integers	115	Transparent Chips, Markers, Gameboard p. 116			✓
G	**LCM Bingo**	Least Common Multiples	37	Special Number Cubes, Markers, Gameboard p. 38			✓
G	**Line Up Four**	Fractions, Decimals, and Percents	98	Transparent Markers, Digit Squares, Gameboard p. 99			✓
S	**Modeling Equations**	Representing Algebraic Equations	126	Transparent Chips, Opaque Tiles, Forms pp. 127–128	✓	✓	
S	**Mystery Factor**	Mental Estimation and Computation	68	Overhead Calculator with Constant Feature, Form p. 69	✓	✓	
S	**Negatives Score**	Modeling Integer Operations	108	Number Cubes, Form p. 109	✓	✓	
G	**Neighboring Integers Count**	Operations with Integers	119	Markers, Digit and Negative Integer Squares, Gameboard p. 120			✓

G = Games S = Sponges

Matrix of Games and Sponges (cont.)

Type	Title	Topic	Page	Materials	Class	Groups	Pairs
G	**Ordered Pathways**	Comparing Percents, Fractions, and Decimals	100	Digit Cards, Form p. 101			✓
S	**Ordering Percents and Fractions**	Comparing Percents and Fractions	89	Digit Squares	✓	✓	
S	**Prime or Composite?**	Prime and Composite Numbers	30	Markers, Paper Clip and Pencil, Form p. 31	✓	✓	
G	**Prime Target**	Prime Numbers	39	Special Number Cubes, Markers, Gameboard p. 40			✓
G	**Products Bingo**	Mental Division and Multiplication	21	Markers, Calculator, Gameboards pp. 22–23			✓
G	**Respond and Travel**	Equivalent Percents and Fractions	95	Markers, Clue Cards p. 97, Gameboard p. 96			✓
G	**Seeking Fractions**	Mentally Computing with Fractions	59	Number Cubes or Digit Squares, Form p. 60			✓
S	**Phrases to Symbols**	Algebraic Representations	129	Teacher Prepared Word Phrases	✓	✓	
G	**Solve and Travel**	Solving Equations	140	Special Number Cubes, Markers, Gameboards pp. 141–142			✓
S	**Subtracting Integers**	Subtracting Integers	106	Transparent Digit and Negative Integer Squares, Transparent Chips, Form p. 107	✓	✓	
S	**Use Your Head**	Mental Estimation and Computation	14	Digit Squares Including Transparent Set, Form p. 15	✓	✓	
S	**What's My Rule?**	Algebraic Rules	130	Prepared Rules, Form p. 131	✓	✓	
S	**What's My Word?**	Percents, Fractions, and Decimals	86	Prepared Clues	✓	✓	

G = Games S = Sponges

Mental Computation

Assumptions Computation operations have previously been taught and reviewed, emphasizing understanding and building operation sense. A variety of mental estimation and computation strategies have been explored and shared by students. Mental computation is promoted regularly and frequently.

Section Overview and Suggestions

Sponges

Compute and Score pp. 12–13

Use Your Head pp. 14–15

These whole-class or small-group warm-ups are open-ended and repeatable. They reinforce the use of all operations and require mental estimation and computation. Both Sponges will ensure greater success with the Games and Independent Activities in this section, especially if you use *Compute and Score* prior to the Game *Compute and Capture.*

Skill Checks

Quick Checks 1-6 pp. 16–18

These provide a way for parents, students, and you to see students' improvement with mental computation. Copies may be cut in half so that each check may be used at a different time. Be sure to have all students respond to STOP, the number sense task, before they solve the ten problems. These Skill Checks are unique because they attempt to promote a mental computation rather than a paper-and-pencil approach.

Games

Compute and Capture p. 19

Draw 4 p. 20

Products Bingo pp. 21–23

These repeatable Games actively involve students in mental computation. *Draw 4* can easily be played as a division game (see *Tips*), while *Products Bingo* provides easier and more challenging gameboard choices.

Independent Activities

Choose and Compute p. 24

Mental Cross-Number Challenge p. 25

Create and Compute p. 26

Each Independent Activity requires students to mentally compute on their own. The open-ended formats of *Choose and Compute* and *Create and Compute* can be duplicated with choices of new digits and problems to provide additional mental-computation practice.

Compute and Score

Topic: Mental Facts and Operations Practice

Object: Use designated digits to make specified amount.

Groups: Whole class or small group

Materials

- *Compute and Score* recording sheet, p. 13
- set of Digit Squares (for each student), p. 147
- set of transparent Digit Squares in an opaque container
- scratch paper

Directions

1. The leader selects, announces, and displays the day's target number. Initially, numbers less than 20 work best.
2. The leader or a student volunteer draws and displays five Digit Squares.
3. Students are challenged to use as many of the digits as possible to form an expression that equals the day's target number. Each digit can be used only once. (If desired, digits can be combined to form 2- or 3-digit numbers.)
4. Students are given time to produce expressions that use as many of the displayed digits as possible.
5. Students receive points according to how many displayed digits they use to yield the day's target number. Students receive 1 point for using 3 digits, 2 points for using 4 digits, and 4 points if they are able to use all 5 digits.

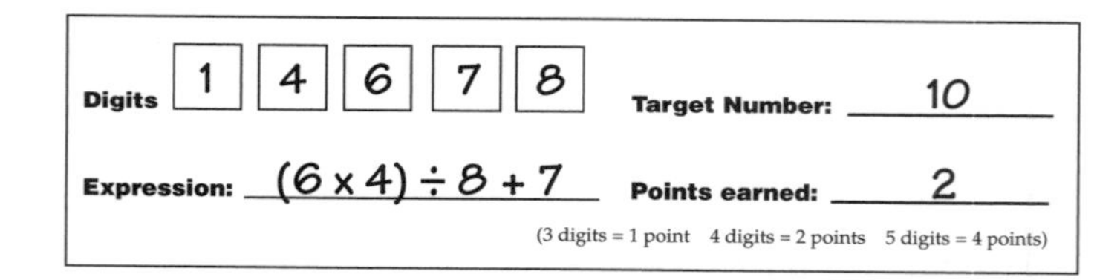

6. After recording their expressions and the points earned, students share and discuss their results.
7. Additional rounds can be played throughout the week with different target numbers and newly drawn digits. Points can be accumulated and totaled at the end of the week.

Tips *Repeated use of this warm-up should produce noticeable improvement in students' confidence and competence. Remember to advance to target numbers beyond 20. Students enjoy working with relevant numbers (day of the month, month number, room number, and so on).*

Making Connections

Promote reflection and make mathematical connections by asking:

- What strategy did you use to create computations that used more than three digits?

Compute and Score

Recording Sheet

Digits 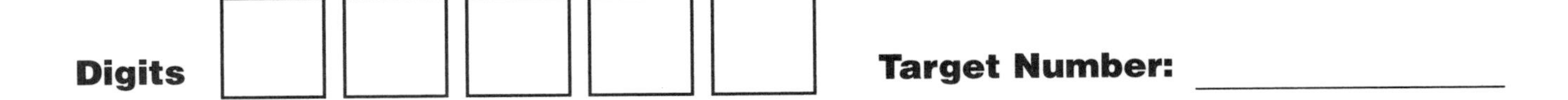 **Target Number:** ____________

Expression: ______________________ **Points earned:** ____________

(3 digits = 1 point 4 digits = 2 points 5 digits = 4 points)

Digits **Target Number:** ____________

Expression: ______________________ **Points earned:** ____________

(3 digits = 1 point 4 digits = 2 points 5 digits = 4 points)

Digits **Target Number:** ____________

Expression: ______________________ **Points earned:** ____________

(3 digits = 1 point 4 digits = 2 points 5 digits = 4 points)

Digits **Target Number:** ____________

Expression: ______________________ **Points earned:** ____________

(3 digits = 1 point 4 digits = 2 points 5 digits = 4 points)

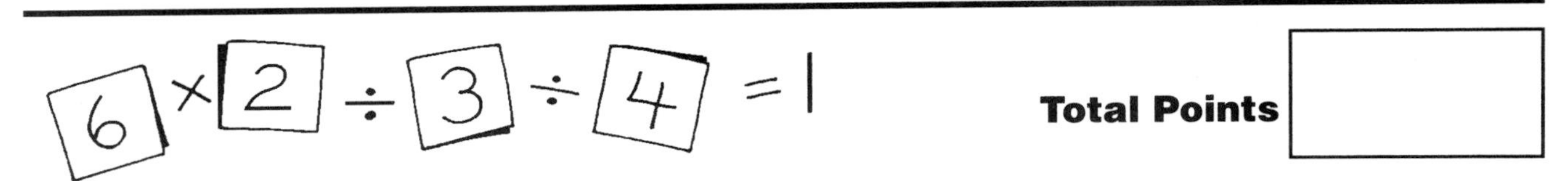

Total Points

Use Your Head

Topic: Mental Estimation and Computation with All Operations

Object: Create problems that meet given criteria.

Groups: Whole class or small group

Tips *Leader can draw an additional Digit Square, and students can discard one drawn number by placing it in a "reject box." To enhance number sense and success rate, announce all drawn digits before students place their Digit Squares.*

Materials

- set of Digit Squares (for each student), p. 147
- set of transparent Digit Squares in an opaque container
- transparency of *Use Your Head,* with forms cut apart, p. 15

Directions

1. The leader selects and displays the problem format (see illustration) which is copied by the students. The leader informs students that a specified number of Digit Squares will be drawn and announced one at a time. Students are informed that Digit Squares are not returned to the container until the end of the round.
2. The leader describes the desired outcome for the problem.

   Example: Create a multiplication problem that produces a product greater than one thousand.

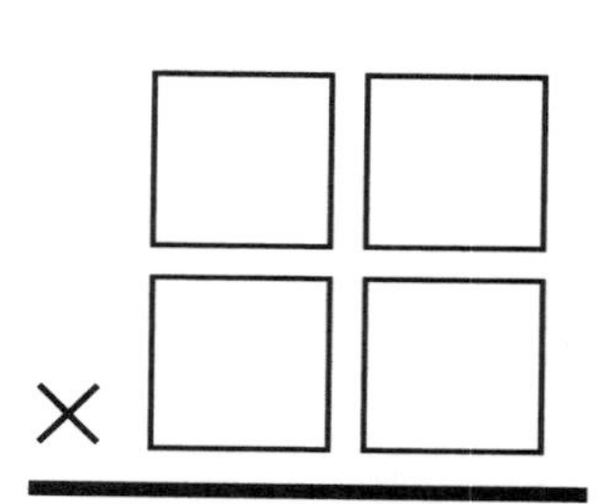

3. The leader draws the first Digit Square and announces the digit. The leader makes sure that each student places the announced digit in one of the squares before drawing the next Digit Square.
4. Students may not change or move placed Digit Squares.
5. This procedure is followed until the required number of Digit Squares is drawn. (Four Digit Squares would be needed for the example above.)
6. After students place all announced Digit Squares, they compute to determine which products have the desired outcome.
7. After students play an additional round with the same outcome, a different desired outcome can be announced.

   Possible outcomes for future rounds: sum between 500 and 700, difference closest to 400, product between 1500 and 2000, quotient less than 200

Making Connections

Promote reflection and make mathematical connections by asking:

- How did you decide where to place certain digits?
- If you were allowed to rearrange your digits, how close could you get to the given criteria?

Use Your Head

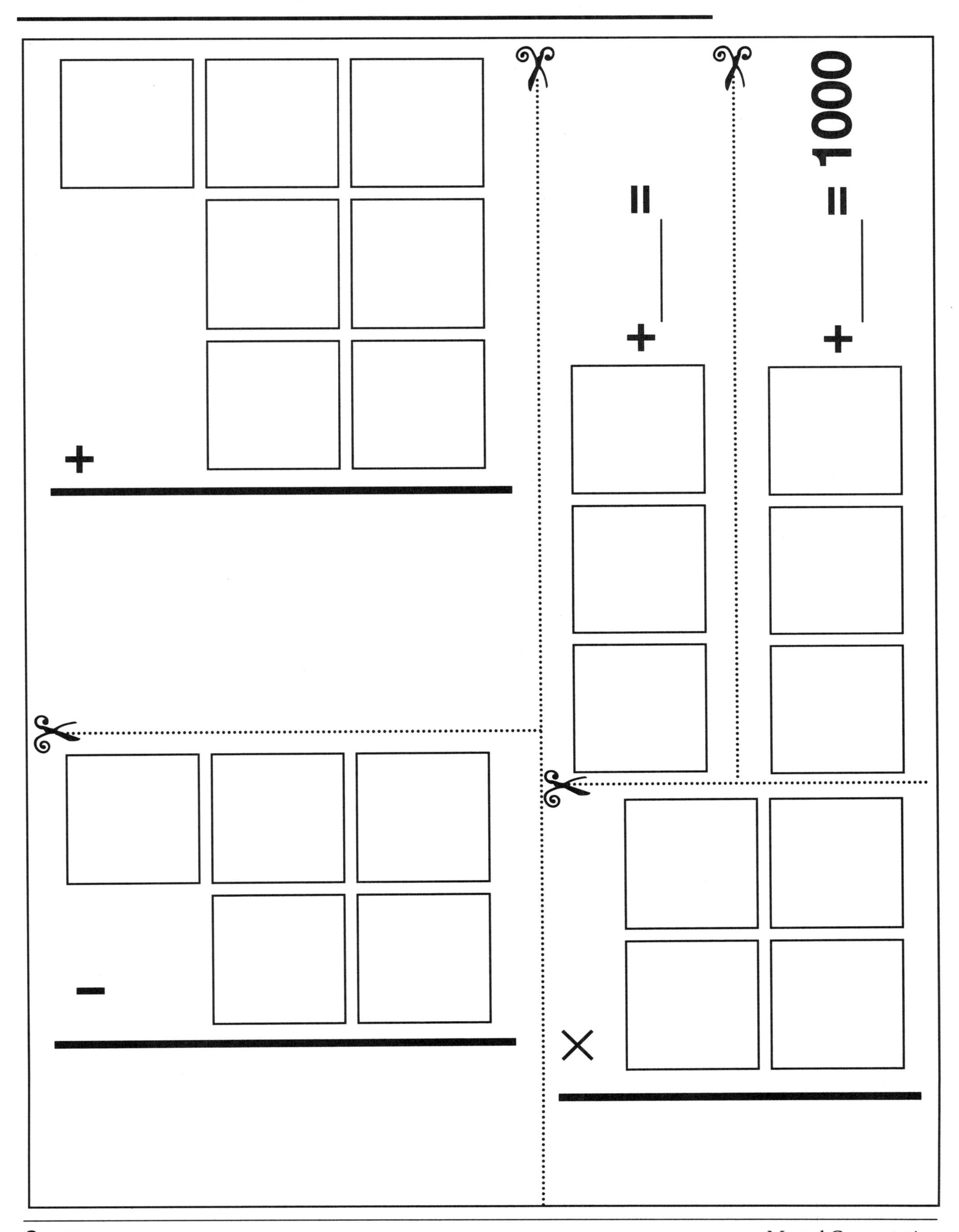

Date ____________ Name ____________________

Quick Checks 1

Don't start yet! Star the problem that may have the greatest answer.

Important: Try to compute each answer mentally, and record only your final answer.

1. $583 - 298 =$ ______ **2.** $673 +$ ______ $= 1000$ **3.** $37 \times 40 =$ ______

4. $14 \times 8 =$ ______ **5.** $152 \div 8 =$ ______ **6.** $70 \times 130 =$ ______

7. $32 + 68 + 145 + 31 =$ ______ **8.** $457 \div 3 =$ ______

9. $6 \times 4 \times 30 \times 5 =$ ______ **10.** $28 \times 25 =$ ______

Use the digits 1, 3, 4, 6, and 7 to write a multiplication problem with the greatest possible product. Describe your strategy.

Date ____________ Name ____________________

Quick Checks 2

Don't start yet! Star two problems that may have answers between 1000 and 4000.

Important: Try to compute each answer mentally, and record only your final answer.

1. $635 - 447 =$ ______ **2.** $1436 +$ ______ $= 2000$ **3.** $23 \times 60 =$ ______

4. $18 \times 8 =$ ______ **5.** $196 \div 7 =$ ______ **6.** $40 \times 170 =$ ______

7. $36 + 73 + 192 + 24 =$ ______ **8.** $438 \div 4 =$ ______

9. $9 \times 3 \times 60 \times 5 =$ ______ **10.** $34 \times 99 =$ ______

Go On Use +, –, ×, or ÷ to make each equation true.

45 ☐ (420 ☐ 6) = 3150 240 ☐ (17 ☐ 9) = 87

Date ______________ Name ______________________

Quick Checks 3

STOP Don't start yet! Star problems that may have odd answers.

Important: Try to compute each answer mentally, and record only your final answer.

1. $718 - 323 =$ ______

2. $543 +$ ______ $= 1000$

3. $28 \times 70 =$ ______

4. $13 \times 8 =$ ______

5. $153 \div 9 =$ ______

6. $60 \times 150 =$ ______

7. $48 + 62 + 178 + 34 =$ ______

8. $428 \div 3 =$ ______

9. $7 \times 4 \times 40 \times 5 =$ ______

10. $32 \times 25 =$ ______

Go On Use the digits 2, 3, 6, and 7 to write a division problem with the greatest whole-number quotient. Describe your strategy.

Date ______________ Name ______________________

Quick Checks 4

STOP Don't start yet! Star three problems that may have answers between 300 and 400.

Important: Try to compute each answer mentally, and record only your final answer.

1. $902 - 576 =$ ______

2. $1683 +$ ______ $= 2000$

3. $25 \times 60 =$ ______

4. $17 \times 6 =$ ______

5. $208 \div 8 =$ ______

6. $70 \times 120 =$ ______

7. $27 + 75 + 183 + 26 =$ ______

8. $546 \div 4 =$ ______

9. $8 \times 4 \times 60 \times 5 =$ ______

10. $29 \times 99 =$ ______

Go On Use the digits 2, 3, 5, 6, and 8 to write a multiplication problem with the smallest possible product. Describe your strategy.

Date ____________ Name ____________

Quick Checks 5

STOP Don't start yet! Star a problem that may have answers that are multiples of 100.

Important: Try to compute each answer mentally, and record only your final answer.

1. $749 - 453 =$ ______ **2.** $467 +$ ______ $= 1000$ **3.** $36 \times 40 =$ ______

4. $16 \times 9 =$ ______ **5.** $126 \div 7 =$ ______ **6.** $40 \times 140 =$ ______

7. $37 + 83 + 167 + 21 =$ ______ **8.** $470 \div 3 =$ ______

9. $8 \times 3 \times 70 \times 5 =$ ______ **10.** $36 \times 25 =$ ______

Go On Use $+$, $-$, $\times$, or $\div$ to make each equation true.

630 □ (35 □ 20) = 42   560 □ (14 □ 8) = 5

Date ____________ Name ____________

Quick Checks 6

Don't start yet! Star two problems that may have 4-digit answers.

Important: Try to compute each answer mentally, and record only your final answer.

1. $811 - 626 =$ ______ **2.** $1572 +$ ______ $= 2000$ **3.** $27 \times 50 =$ ______

4. $19 \times 7 =$ ______ **5.** $243 \div 9 =$ ______ **6.** $60 \times 160 =$ ______

7. $29 + 77 + 159 + 32 =$ ______ **8.** $573 \div 4 =$ ______

9. $9 \times 4 \times 50 \times 5 =$ ______ **10.** $27 \times 99 =$ ______

Go On Use the digits 5, 6, 7, and 9 to write a division problem with the smallest whole-number quotient. Describe your strategy.

Compute and Capture

Topic: Mental Facts and Operations Practice

Object: Create varied computations to yield displayed target numbers.

Groups: 2 players

Materials for each group

- 3 sets of Digit Cards, p. 146
- consumable paper strips

Tips Students less proficient with their facts will benefit by initially playing as pair players. Students will gain additional practice if 4-by-4 arrays are displayed for target numbers. Cards will need to be remixed for the second round.

Directions

1. After mixing Digit Cards, display cards as illustrated. The cards in the 3-by-3 array are the "target numbers." The five cards above are the "digit choices." Stack the remaining cards facedown for the second round.
2. Players use four or five of the digit choices and any operations to create expressions equal to the displayed target numbers. Digits can be combined to form 2- or 3-digit numbers to use in the expressions.
3. Players are given adequate time to create as many expressions as possible for the target numbers. Once players find a solution, they record it on paper strips.
4. When both are ready, the players alternate turns sharing and displaying expressions for a specified target number. It is important that written expressions follow the standard order of operations.
5. After both players agree that an expression is correct, the player who wrote it captures the Digit Card with the target number. When duplicate target numbers are displayed, a player must have a different expression in order to capture the duplicate card. (It may be helpful to refer to the previously recorded paper strip.)

   Example: Each player captures a 9 by displaying the following expressions: $(2 \times 8) - (5 + 2)$ and $(8 \times 5) \div (2 \times 2) - 1$
6. Players alternate turns until they have shared all their possibilities.
7. Gather the remaining displayed Digit Cards and place them in a discard stack. Play a second round by arranging the cards in the unused stack to display five new digit choices and a new array of target numbers. The player who shared second in the first round will share first in this round.
8. The player who has captured more Digit Cards is the winner.

Digit Choices

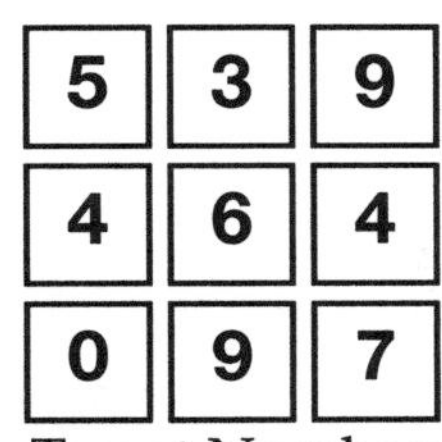

Target Numbers

Making Connections

Promote reflection and make mathematical connections by asking:

- Which digits were desirable as digit choices? Why?
- What strategies would you recommend to friends for future rounds?

Draw 4

Topic: Mental Multiplication and Operation Sense

Object: Given a product and partial-factor clues, identify the hidden digit.

Groups: 2–3 players

Materials

- Digit Squares in opaque container, p. 147
- scratch paper

Directions

1. The leader displays the problem format (see illustration) which is copied by each player. Players erect a barrier to hide their work area from other players.
2. Players draw four Digit Squares from their separate containers, keeping the digits hidden from the other players.
3. After players place each of the four Digit Squares in a cell, they accurately and carefully compute the product, which is then recorded below the problem format. Each player turns over one of the four Digit Squares to hide its identify.
4. When all players have completed Step 3, each player passes his or her problem to the player to the left.
5. Using the displayed product and three of the four digits as clues, each player tries to quickly identify the hidden digit. After the players record their predicted hidden digits, they announce, "Ready."
6. When all the players have called "Ready," they use their predictions to compute the product with pencil and paper. The hidden Digit Square can be turned over to further verify the match.
7. If there is a correct match, a player receives 1 point. The player who first correctly identifies and completes the task receives an additional point.
8. If there is not a match and the product is found to be in error, the discovering player receives a point.
9. Players are encouraged to play at least three rounds before announcing a winner—the player with the highest accumulated total.

Tips *Adapt this game to division by hiding one digit of the divisor or dividend and displaying the entire quotient. The problem format can be easily altered for either version to better accommodate students' skill level.*

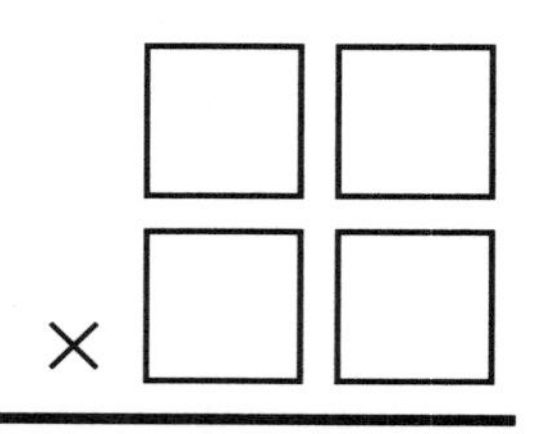

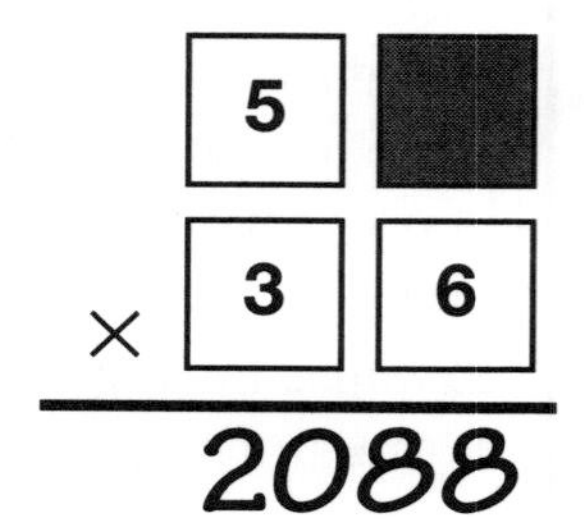

Making Connections

Promote reflection and make mathematical connections by asking:

- What clues helped you identify the hidden digit?

Products Bingo

Topic: Mental Division and Multiplication

Object: Cover four products in a row.

Groups: 2 players

Tip *Consider using the more challenging factor choices on* Products Bingo B *gameboard, p. 23.*

Materials for each group

- *Products Bingo A* gameboard, p. 22
- 2 kinds of markers
- calculator

Directions

1. The first player selects and announces two factors from the "Choices" box. After announcing the factor choices, the player may use a calculator or pencil and paper to compute the product. The first player covers the resulting product on the gameboard with his or her marker. If there are two of the same product on the gameboard, only one product may be covered on a turn.
2. The other player selects and announces two factors and then computes the product. If the resulting product is on the gameboard and not covered, that player covers it with his or her marker.
3. Players continue alternating turns.
4. The first player to have four markers in a row horizontally, vertically, or diagonally wins.

Choices

13	19	24	28	32	35	47	50

1316	416	455	1200	672
608	1400	1645	312	893
665	768	950	1600	1120
532	1128	247	840	1645
456	980	1750	1316	896
2350	611	364	650	1504

Making Connections

Promote reflection and make mathematical connections by asking:

- What strategy helped you place your markers in a row?

Products Bingo A

Choices

13	19	24	28	32	35	47	50

32 × ? = 1600

1316	416	455	1200	672
608	1400	1645	312	893
665	768	950	1600	1120
532	1128	247	840	1645
456	980	1750	1316	896
2350	611	364	650	1504

Products Bingo B

Choices

24	31	58	66	75	89	97	113

2328	1798	5874	1800	10,057
3828	5626	2759	6554	4350
3503	8475	8633	744	6675
4950	2046	10,961	3007	5874
2712	7275	2136	1392	7458
5162	2325	6402	5626	1584

Date ____________________ Name ________________________________

Choose and Compute

Factor Choices: 18, 27, 35, 48, 56, 60, 72, 99

Choose two of the factors above that will give the products below.

1. $\times$ ___ = 2520

2. $\times$ ___ = 1512

3. $\times$ ___ = 5544

4. $\times$ ___ = 2880

Divisor Choices: 17, 39, 45, 84, 91
Quotient Choices: 23, 36, 57, 78

Choose one divisor and one quotient above to complete each division problem.

5. $\overline{)2223}$

6. $\overline{)3276}$

7. $\overline{)1932}$

8. $\overline{)1326}$

9. $\overline{)2565}$

10. $\overline{)3510}$

Add multiplication and division signs to make these equations true. A sample is given.

1 2 3 4 5 = 80

1 2 ÷ 3 × 4 × 5 = 80

11. 3 6 0 4 4 5 = 4050

12. 8 2 3 5 5 7 = 82

13. 7 2 4 1 3 4 = 2353

14. 2 2 9 6 4 1 3 = 168

15. 2 4 9 3 9 3 8 = 10,526

Date ____________________ Name ______________________________

Mental Cross-Number Challenge

First try to mentally compute each problem without pencil and paper. Record your mental computation strategies for two problems. Be prepared to share your strategies with the class.

Across:

1. $4760 - 99$
4. $6608 \div 8$
6. 317×9
8. $870 \div 30$
10. $1000 - 290$
11. $11{,}000 - 38$
12. $441 \div 9$
13. $5200 - 4583$
15. $(9270 \div 30) + 206$
17. $1632 + 386 - 1480$
18. $(1165 \div 5) - 30$
19. $2100 \div 50$
21. $3500 + 1279 + 4300$
23. $2357 - 1599$
24. $24{,}000 \div 800$

1.		2.	3.		4.		5.
		6.		7.			
8.	9.		10.				
11.						12.	
				13.	14.		
15.		16.			17.		
	18.				19.		
20.				21.			22.
23.				24.			

Down:

1. $4570 - 349$
2. $3100 \div 50$
3. 7×268
4. $83{,}000 \div 100$
5. $10{,}000 - 3661$
7. $15{,}378 \div 3$
9. $11{,}289 \times 8$
12. $11{,}589 + 35{,}738$
14. 77×200
16. 313×16
20. $(1260 \div 30) - 15$
21. $(1395 \div 5) - 186$
22. $5237 - 5138$

Trivia for 3 Down:
The first French fries were made in Belgium in this year. How long have they been around?

Date ____________ Name ____________

Create and Compute

Arrange three of the four digits 2, 3, 5, and 8 to create and solve a problem for each goal given below.

1.

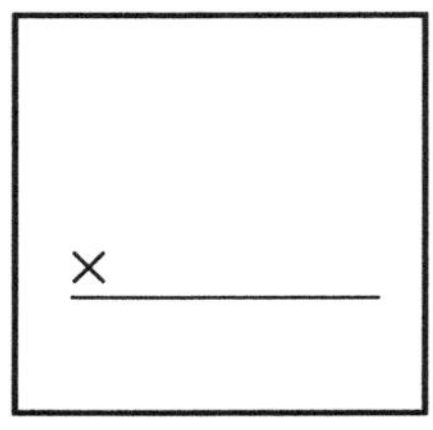

A product between 150 and 200

2.

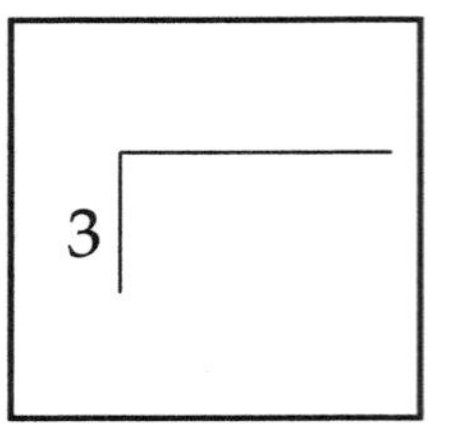

A quotient greater than 200 without a remainder when divided by 3

3.

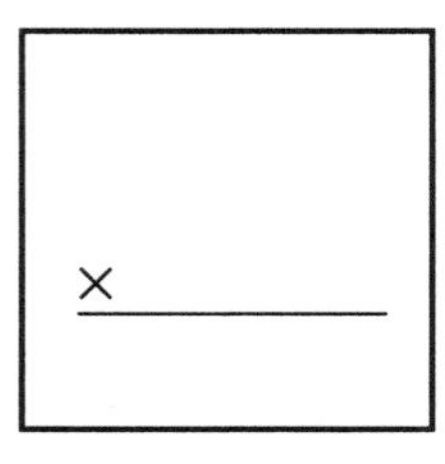

An even product greater than 400

4.

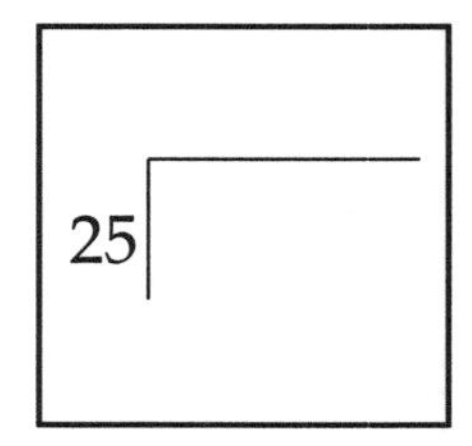

A quotient without a remainder when divided by 25

Arrange all four of the digits 2, 4, 7, and 8 to create and solve a problem for each goal given below.

5.

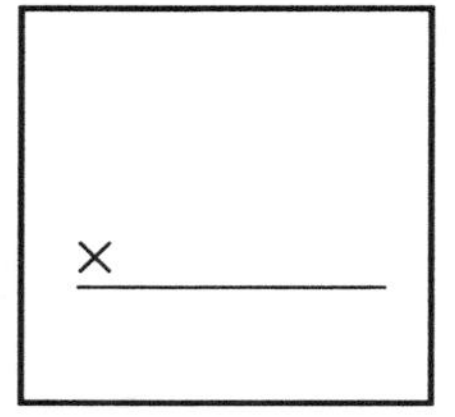

A product between 3000 and 3500

6.

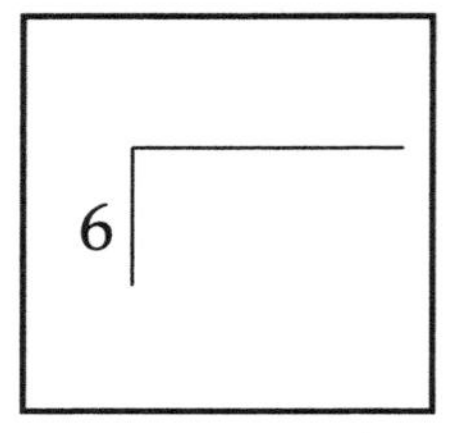

A quotient greater than 700 without a remainder when divided by 6

7.

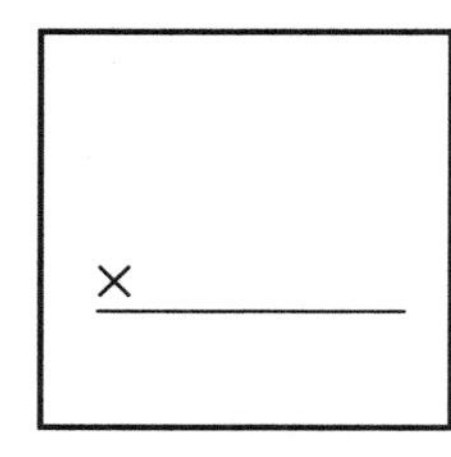

An even product between 3500 and 4000

8.

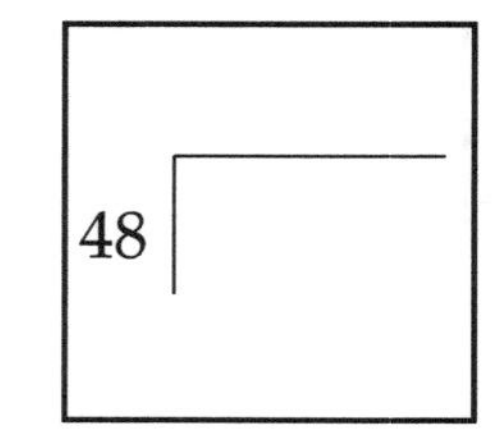

A quotient without a remainder when divided by 48

Arrange all four of the digits 2, 6, 7, and 9 to create and solve a problem for each goal given below.

9.

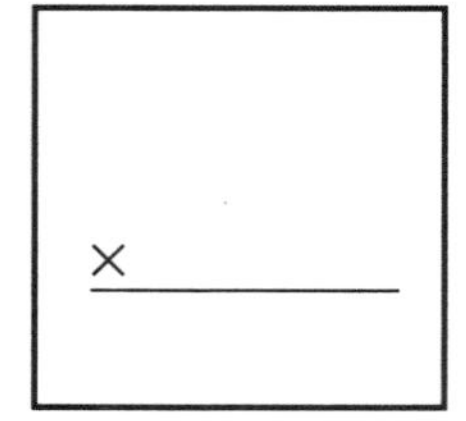

A product between 2000 and 2200

10.

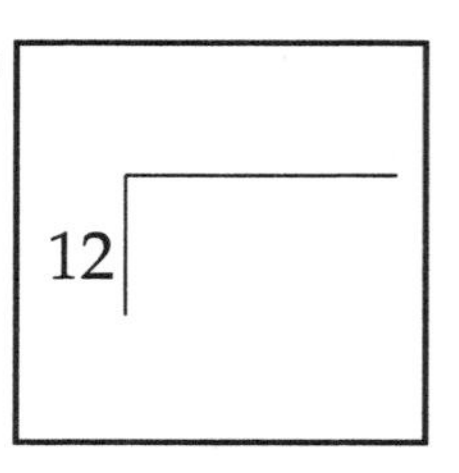

A quotient between 750 and 800 without a remainder when divided by 12

11.

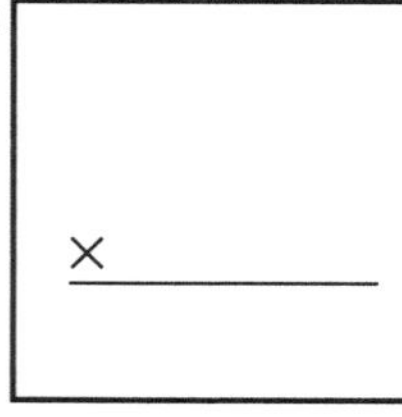

An odd product less than 2000

12.

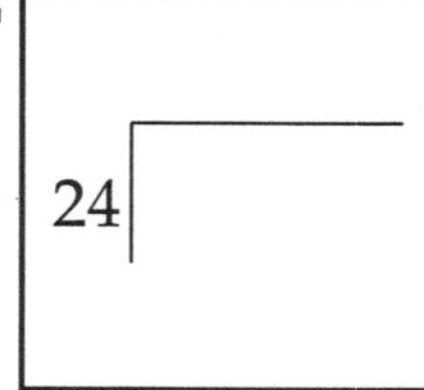

A quotient without a remainder when divided by 24

Number Theory

Assumptions Number theory topics such as primes, common factors, and least common multiples have previously been taught, emphasizing understanding and building number sense. A variety of visual and symbolic models, such as factor trees, have deepened understanding and promoted mathematical connections.

Section Overview and Suggestions

Sponges

Greatest Common Factor Bingo pp. 28–29

Prime or Composite? pp. 30–31

Common Factors Match pp. 32–33

These whole-class or small-group warm-ups are open-ended and repeatable. They promote mental computation as number concepts are reinforced. Use of these Sponges will ensure greater success with the Games and Independent Activities in this section, especially if you use *Prime or Composite?* prior to the *Prime Target* Game. All three Sponges can be easily adapted into games.

Skill Checks

Number Explorations 1–6 pp. 34–36

The Skill Checks provide a way for parents, students, and you to see students' improvement with number theory topics. Copies may be cut in half so that each check may be used at a different time. Remember to have all students respond to STOP, the number-sense task, before starting.

Games

LCM Bingo pp. 37–38

Prime Target pp. 39–40

Factor Find pp. 41–42

These open-ended and repeatable Games actively involve students in mental computation as they practice finding primes, multiples, and factors. With the repeated use of *LCM Bingo* and *Factor Find,* students will enhance their reasoning abilities as they develop more effective game strategies.

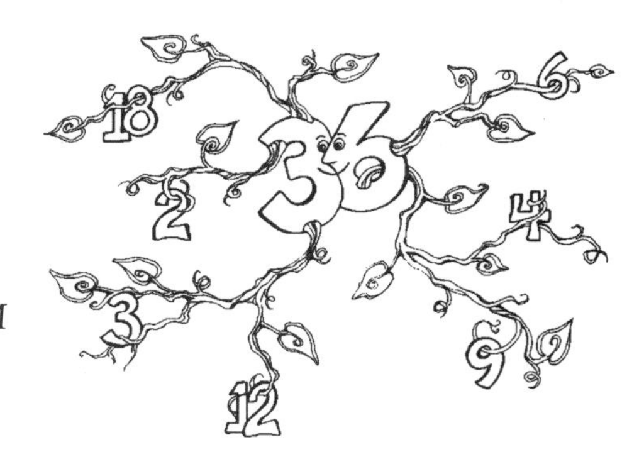

Independent Activities

Finding Primes p. 43

Common Factors Practice pp. 44–45

Using LCM and GCF to Check Multiplication p. 46

Each Independent Activity requires students to practice finding common factors, primes, least common multiples, and greatest common factors. All the formats can be duplicated with new choices and problems to provide additional practice. Use of *Common Factors Match* (Sponge) will ensure greater success with *Common Factors Practice.* It is important for students to succeed with both before attempting *Common Factors Practice Challenge.*

Greatest Common Factor Bingo

Topic: Greatest Common Factor

Object: Cover three numbers in a row.

Groups: Whole class or small group

Materials

- transparency of *Greatest Common Factor Bingo* playing board, p. 29
- 2 kinds of markers
- pencil and 2 paper clips for spinners

Tips *Require each team to have different team members create and display the supporting evidence in the form of factor trees. This warm-up can be easily adapted into a game for two players or pair players.*

Directions

1. Divide the class or group into two teams. Team A uses the pencil and paper clips with the two spinners to identify two composite numbers. Team A announces the two spun numbers.
2. Team B announces the greatest common factor for the two numbers spun and diagrams the two corresponding factor trees to provide proof. (The greatest common factor can be found by multiplying one set of the factors common to both bottom rows of the factor trees.)

   Example: 24 and 60 are spun. Team B states, "$2 \times 2 \times 3 = 12$, so twelve is the greatest common factor" and displays a factor tree proof.

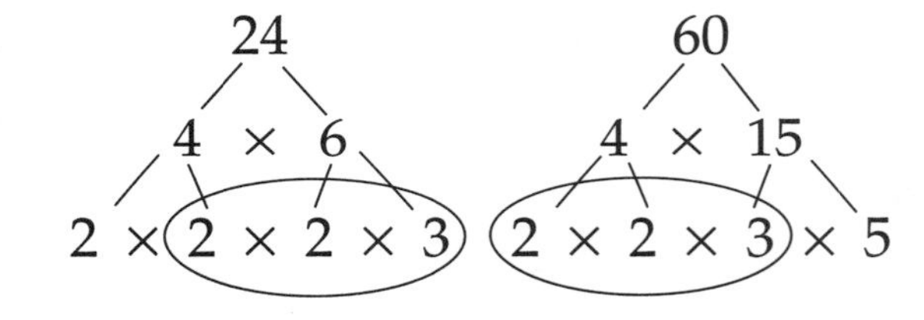

3. Once Team A agrees with Team B's response, Team B covers the resulting greatest common factor. If the greatest common factor appears twice on the playing board, only one may be covered on a turn.
4. Team B spins and announces the two spun numbers. After Team A states the greatest common factor for the two numbers and provides correct factor trees, Team A covers a corresponding cell on the playing board.
5. If all the cells of the resulting greatest common factor are already covered, the team may replace an opponent's marker with one of its own.
6. This procedure and alternating of turns continues until one team positions three markers in a row horizontally, vertically, or diagonally.

Making Connections

Promote reflection and make mathematical connections by asking:

- What approach helped you quickly identify the greatest common factor?

Greatest Common Factor Bingo

4	10	6	4
12	8	4	16
4	8	12	4
8	2	20	6

Prime or Composite?

Topic: Prime and Composite Numbers

Object: Cover three cells in a row.

Groups: Whole class or small group

Materials

- transparency of *Prime or Composite?* playing board, p. 31
- 2 kinds of markers
- pencil and 2 paper clips for spinners

Directions

1. Divide the class or group into two teams. Team A uses the pencil and paper clips with the two spinners to identify a 2-digit number. Team A announces the sum of the two spun numbers.
2. Team B states whether the announced number is a prime or composite number and provides supporting evidence.

   Example: "Ninety-one is a composite number because it is divisible by 7 or $7 \times 13 = 91$."
3. Once Team A agrees with Team B's response, Team B is allowed to cover one of the Cs or Ps on the playing board. Each team uses its own type of marker.

   Example: Based on the response above, Team B would cover a **C** since 91 is a composite number.
4. Team B spins and announces the new 2-digit number. After Team A states whether the spun number is a prime or composite number and provides supporting evidence, Team A covers a **C** or **P** cell on the playing board.
5. This procedure and alternating of turns continues until one team positions three markers in a row horizontally, vertically, or diagonally.

__Tips__ Require each team to have different team members create and display the supporting evidence by identifying factors or showing that the number is not divisible by 3, 7, or 9. This warm-up can be easily adapted into a game for two players or pair players.

Making Connections

Promote reflection and make mathematical connections by asking:

- What helped your team easily determine whether an announced number was a prime or composite number?
- Which numbers are more likely to result with these spinners, prime or composite? Please explain.

Prime or Composite?

C	P	C	P
P	C	P	C
C	P	C	P
P	C	P	C

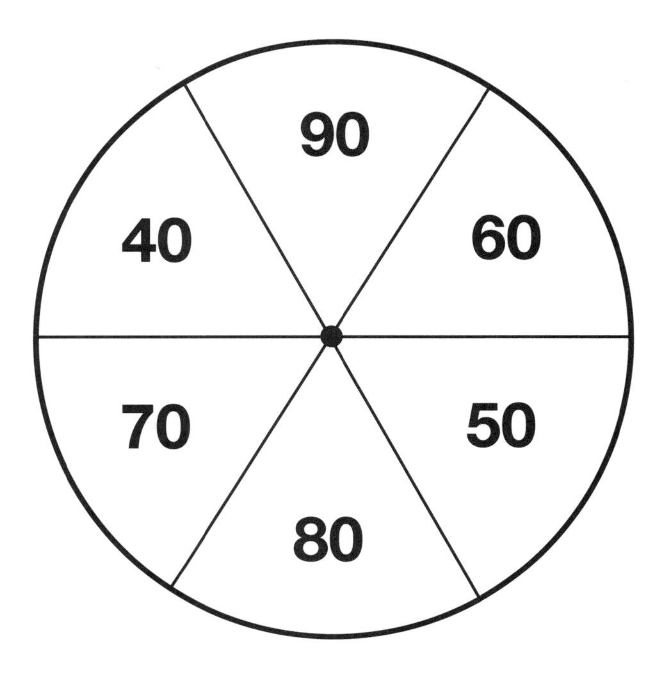

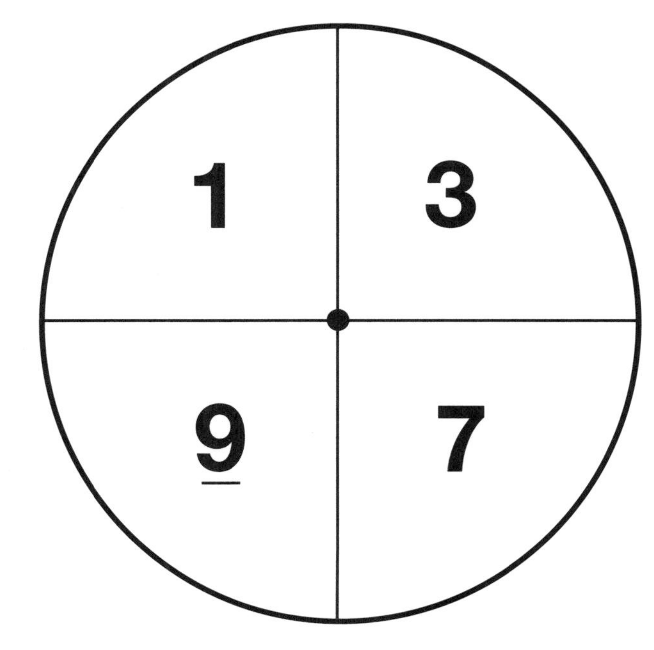

Common Factors Match

Topic: Common Factors

Object: Place available numbers so every number shares a common factor with every adjacent number.

Groups: Whole class or small group

Materials

- transparency of *Common Factors Match* recording grid, p. 33
- overhead pens

Tips *After a round is tried, students can partner up to attempt the activity as pairs. This warm-up activity can be simplified by using a 4-by-4 or 3-by-3 grid and fewer numbers. (See pages 44 or 45.)*

Directions

1. The class is divided into two to four groups. Groups take turns selecting and placing numbers. One volunteer from the first group selects one number to place in the recording grid.

   Example: "We choose 9. Please record it in the cell located in the second row and third column." 9 is crossed out after it is recorded in the identified cell.
2. One volunteer from the second group selects a number and the place to record it in the recording grid. This second number must share a common factor with the previously placed number and adjoin a previously recorded number.
3. A volunteer from the next group follows the same procedure, making sure that the number placed shares at least one common factor with the numbers in all the adjacent cells.
4. Groups continue to take turns selecting and cautiously placing numbers.
5. If a number is recorded that does not share a common factor with each adjacent cell, the number is removed and again becomes a number choice.
6. The goal is for the groups to collaborate in order to successfully place all 25 numbers, so every number shares a common factor with each adjacent cell horizontally, vertically, and diagonally.

Numbers to Place:			2	3	4	5
6	8	~~9~~	10	12	15	16
18	20	24	25	~~27~~	30	32
~~36~~	45	48	50	60	64	~~72~~

	27	9		72
			36	

Making Connections

Promote reflection and make mathematical connections by asking:

- Which numbers are good beginning numbers? Explain your reasoning.
- What strategies helped the class fill all, or almost all, of the cells?

Common Factors Match

Numbers to Place:			**2**	**3**	**4**	**5**
6	**8**	**9**	**10**	**12**	**15**	**16**
18	**20**	**24**	**25**	**27**	**30**	**32**
36	**45**	**48**	**50**	**60**	**64**	**72**

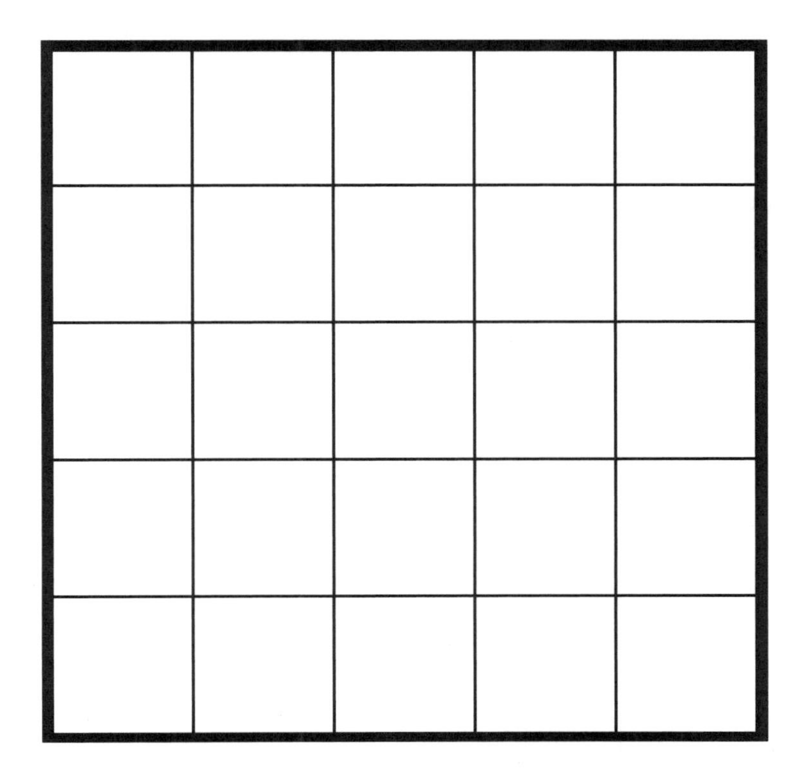

Date ____________ Name ____________

Number Explorations 1

STOP Don't start yet! Star a problem that may have an answer that is a multiple of 12.

Find the least common multiple and the greatest common factor for each pair of numbers.

8 and 20 **1.** LCM ______ **2.** GCF ______ 28 and 36 **3.** LCM ______ **4.** GCF ______

5. List all prime numbers between 30 and 45. ____________

6. Name a number that is a multiple of 4 and 9 and greater than 50. ____________

7. Name a multiple of 6 and 4 that is between 30 and 50. ____________

8. Express 54 as a product of primes. ____________

Fill the Venn diagrams with all the numbers from 2, 5, 9, 12, 16, 17, 21, and 25 that fit.

9. primes multiples of 4

10. factors of 60

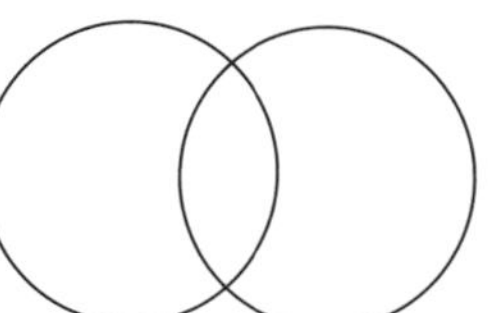

composite numbers

Go On Create a factor tree for 432.

Date ____________ Name ____________

Number Explorations 2

STOP Don't start yet! Star a problem that may have a 1-digit answer greater than 5.

Find the least common multiple and the greatest common factor for each pair of numbers.

15 and 20 **1.** LCM ______ **2.** GCF ______ 27 and 18 **3.** LCM ______ **4.** GCF ______

5. List all prime numbers between 55 and 70. ____________

6. Name a number that is a multiple of 3 and 8 and greater than 80. ____________

7. Name a multiple of 7 and 5 that is between 100 and 150. ____________

8. Express 120 as a product of primes. ____________

Fill the Venn diagrams with all the numbers from 2, 4, 6, 9, 12, 32, 40, and 64 that fit.

9. multiples of 6 multiples of 4

10. factors of 72

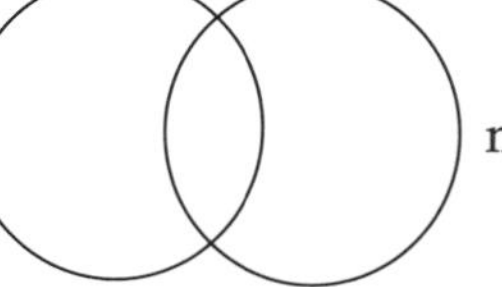

multiples of 8

Go On I am a perfect square, the sum of three different primes, and less than 40. What number am I?

Date ____________ Name ____________

Number Explorations 3

STOP Don't start yet! Star a problem that may have an answer that is a multiple of 10.

Find the least common multiple and the greatest common factor for each pair of numbers.

10 and 12 **1.** LCM ______ **2.** GCF ______ 9 and 24 **3.** LCM ______ **4.** GCF ______

5. List all prime numbers between 40 and 65. ____________

6. Name a number that is a multiple of 5 and 8 and less than 60. ____________

7. Name a multiple of 8 and 12 that is between 20 and 50. ____________

8. Express 100 as a product of primes. ____________

Fill the Venn diagrams with all the numbers from 2, 3, 6, 9, 18, 23, 36, and 41 that fit.

9.

multiples of 6 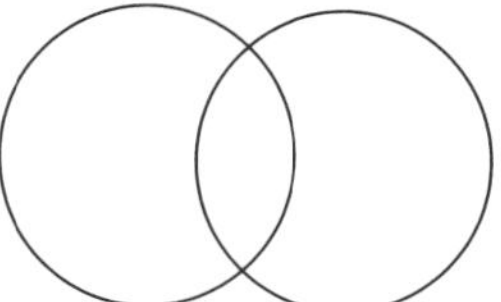 factors of 54

10.

composite numbers 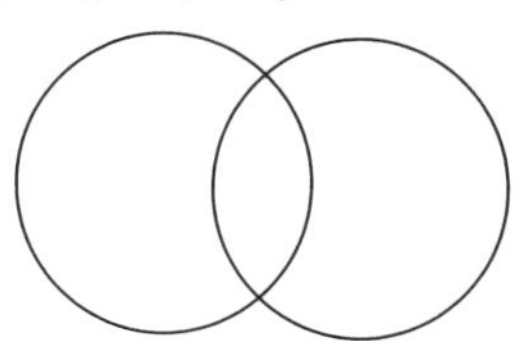 factors of 18

Go On List all the primes between 125 and 175.

Date ____________ Name ____________

Number Explorations 4

STOP Don't start yet! Star a problem that may have an answer greater than 85.

Find the least common multiple and the greatest common factor for each pair of numbers.

10 and 18 **1.** LCM ______ **2.** GCF ______ 15 and 18 **3.** LCM ______ **4.** GCF ______

5. List all prime numbers between 80 and 95. ____________

6. Name a number that is a multiple of 6 and 8 and less than 40. ____________

7. Name a multiple of 3 and 8 that is between 40 and 60. ____________

8. Express 80 as a product of primes. ____________

Fill the Venn diagrams with all the numbers from 2, 6, 9, 12, 15, 17, 27, and 31 that fit.

9.

multiples of 3 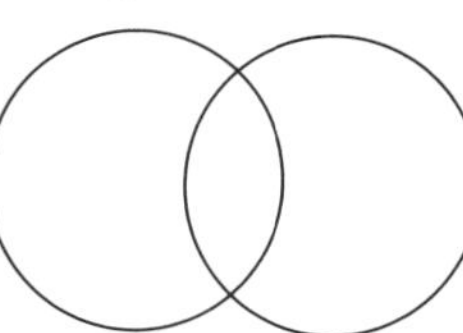 multiples of 5

10.

factors of 36 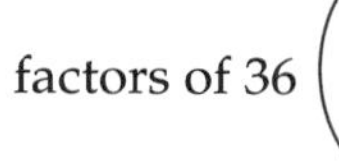 primes

Go On Create a factor tree for 504.

Date ____________ Name ____________

Number Explorations 5

STOP Don't start yet! Star a problem that may have an answer that is a composite number less than 40.

Find the least common multiple and the greatest common factor for each pair of numbers.

8 and 12 **1.** LCM ______ **2.** GCF ______ 12 and 18 **3.** LCM ______ **4.** GCF ______

5. List all prime numbers between 70 and 80. ____________

6. Name a number that is a multiple of 2 and 6 and greater than 80. ____________

7. Name a multiple of 7 and 4 that is between 100 and 130. ____________

8. Express 64 as a product of primes. ____________

Fill the Venn diagrams with all the numbers from 2, 6, 7, 12, 28, 31, 35, and 42 that fit.

9. primes composites

10. multiples of 7 factors of 24

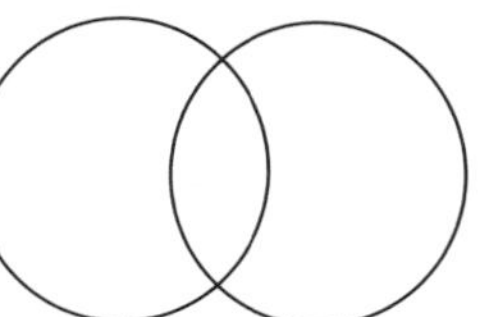

Go On The product of two consecutive page numbers is 3422. What are they?

Date ____________ Name ____________

Number Explorations 6

STOP Don't start yet! Star a problem that may have an answer less than 5.

Find the least common multiple and the greatest common factor for each pair of numbers.

6 and 10 **1.** LCM ______ **2.** GCF ______ 12 and 20 **3.** LCM ______ **4.** GCF ______

5. List all prime numbers between 45 and 55. ____________

6. Name a number that is a multiple of 4 and 6 and less than 50. ____________

7. Name a multiple of 3 and 7 that is greater than 40. ____________

8. Express 110 as a product of primes. ____________

Fill the Venn diagrams with all the numbers from 3, 9, 27, 37, 40, 45, 67, and 72 that fit.

9. multiples of 9 factors of 81

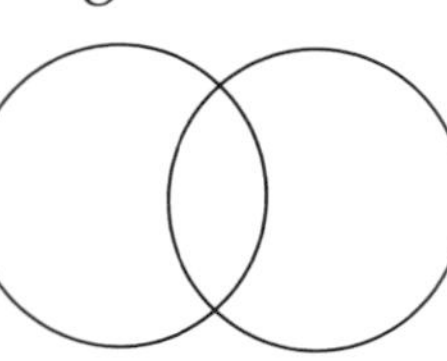

10. multiples of 5 primes

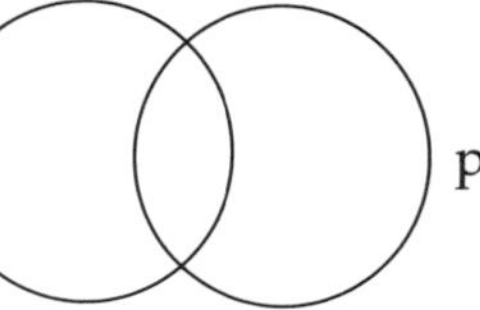

Go On List all the primes between 175 and 225.

LCM Bingo

Topic: Least Common Multiple

Object: Cover four numbers in a row.

Groups: 2 or 3 players

Tip *Play the game without allowing the sharing of cells and compare the results.*

Materials for each group

- *LCM Bingo* gameboard, p. 38
- 2 Number Cubes or 2 spinners (1–6 and 3–8), pp. 149 and 150
- different kind of markers for each player

Directions

1. The first player rolls the number cubes (or spins the spinners) and determines the least common multiple of the two displayed numbers. The player locates and covers the resulting least common multiple on the gameboard. If there are two of the same least common multiple on the gameboard, a player may cover only one on a turn.
2. The second player rolls the cubes, determines the least common multiple, and covers the resulting least common multiple. If the resulting least common multiple is already covered by an opponent's marker, the player may also cover that number. Players are allowed to share cells.
3. If a player rolls a combination that is already covered by one of his or her own markers, the player does not have a play for that turn.
4. Players take turns and continue to roll, identify the least common multiple, and cover one least common multiple.
5. The first player to have four markers in a row horizontally, vertically, or diagonally wins.

Making Connections

Promote reflection and make mathematical connections by asking:

- How would the game be different if the gameboard cells could not be shared?
- What approach helped you easily identify the least common multiple?

LCM Bingo

8	12	5	6	4
12	3	14	7	28
6	40	21	12	20
24	8	30	42	6
12	15	4	35	10

Prime Target

Tip *If hints for locating the prime numbers seem needed, place dots in the margin of each row to indicate the number of primes in that row (from the top down: 4, 4, 2, 2, 3, 2, 2, 3, 2, 1).*

Topic: Prime Numbers

Object: Collect 13 markers by computing randomly generated numbers to yield prime numbers.

Groups: 2 players

Materials for each group

- *Prime Target* gameboard, p. 40
- 25 markers
- 3 Number Cubes (1–6), p. 149
- 1 special Number Cube (4–9), p. 149

Directions

1. Players collaborate to identify the 25 prime numbers found on the Prime Target gameboard and cover these numbers with markers.
2. The first player rolls all four number cubes and uses two or more of the displayed numbers to create an expression whose value is a prime number. The player may form 2- or 3-digit numbers from the rolled digits.

   Example: 3, 3, 5, and 6 are rolled. Player announces, "$6 \times (3 + 3) - 5$ (for 31)." Other examples are $5 + 6$ (for 11), and $56 - 33$ (for 23).
3. After a player creates and shares an expression equal to a prime number, the player removes and keeps the marker that covered the prime number.
4. The second player rolls the four number cubes and follows the same steps. It is important that each player announce his or her expression before removing the covering marker.
5. If a player cannot create an expression equal to a covered prime number, he or she loses that turn.
6. The first player to collect 13 markers wins.

Making Connections

Promote reflection and make mathematical connections by asking:

- What helped you easily locate the 25 prime numbers?
- For which prime numbers was it difficult to create an expression?

Prime Target

1	2	3	4	5	6	7	8	9	10
11	12	13	14	15	16	17	18	19	20
21	22	23	24	25	26	27	28	29	30
31	32	33	34	35	36	37	38	39	40
41	42	43	44	45	46	47	48	49	50
51	52	53	54	55	56	57	58	59	60
61	62	63	64	65	66	67	68	69	70
71	72	73	74	75	76	77	78	79	80
81	82	83	84	85	86	87	88	89	90
91	92	93	94	95	96	97	98	99	100

Factor Find

Topic: Factors and Multiples

Object: Identify high-value multiples and factors to gain the most points.

Groups: 2 players or pair players

Materials for each group

- *Factor Find* gameboard, p. 42
- different kind of markers for each player
- calculator (*optional*)

Directions

1. The first player selects and covers a multiple that has an uncovered factor on the board.
2. The second player covers one factor of that multiple and then covers a new multiple with at least one uncovered factor on the gameboard.
3. The first player covers one factor of the second player's multiple and then covers a new multiple.

   Example: Player 1 covers 40 as a multiple. Player 2 covers 20 (as a factor of 40) and covers 45 as a new multiple. Player 1 covers 15 (as a factor of 45) and covers 35 as a new multiple.

1	2	3	4	5	6	7	8	9
10	11	12	13	14	15	16	17	18
19	20	21	22	23	24	25	26	27
28	29	30	31	32	33	34	35	36
37	38	39	40	41	42	43	44	45

4. This process continues with players alternating turns until there is no uncovered multiple with a factor left on the board.
5. If a player covers an incorrect factor or multiple, the player is allowed to replay.
6. If one player thinks the game is over and the other player finds a remaining play, the discovering player covers the multiple and allows the opponent to cover the remaining factor.
7. Players figure their scores by totaling the value of all the covered numbers. The five prime numbers enclosed by hexagons are worth twice their numerical value. Players may use calculators. The player with the higher score wins.

Tips *You may want to have students understand from the beginning how scores are determined (see direction 7). For a challenge, allow the opposing players to cover one or two factors for each multiple covered and discuss how game strategies change. If desired, make a gameboard that extends the numbers to 80.*

Making Connections

Promote reflection and make mathematical connections by asking:

- What are some good first plays? Why?
- What strategy helped you get a high score?

Factor Find

1	2	3	4	5	6	7	8	9
10	11	12	13	14	15	16	17	18
19	20	21	22	23	24	25	26	27
28	29	30	31	32	33	34	35	36
37	38	39	40	41	42	43	44	45

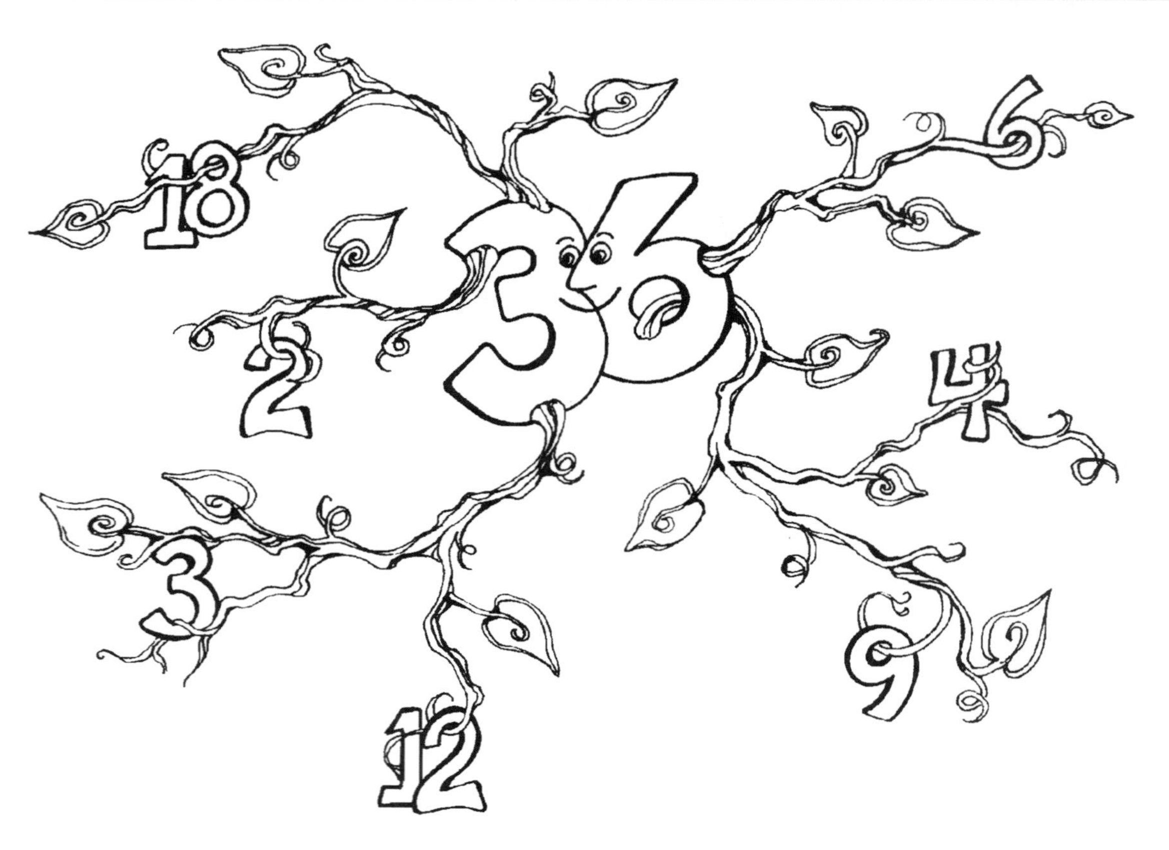

Date ____________ Name ____________

Finding Primes

Identify two prime numbers whose sum is the given number. The first problem has been done for you.

Example:

1. 3 + 7 = 10 **2.** ___ + ___ = 20 **3.** ___ + ___ = 30

4. ___ + ___ = 66 **5.** ___ + ___ = 48 **6.** ___ + ___ = 32

7. ___ + ___ = 44 **8.** ___ + ___ = 112 **9.** ___ + ___ = 82

10. ___ + ___ = 84 **11.** ___ + ___ = 102 **12.** ___ + ___ = 100

Find the prime factors for each composite number by completing a factor tree. The first factor tree has been done for you.

Example:

13. 45

9 × 5

3 × 3 × 5

14. 56

×

15. 72

×

16. 80

×

17. 100

×

18. 120

×

19. 200

×

20. 156

×

21. 126

×

Date ____________________ Name ______________________________

Common Factors Practice

Carefully write the boxed numbers in each grid so that every number placed shares a common factor with every number in an adjacent cell horizontally, vertically, and diagonally. The first grid has been started for you.

3 ~~6~~ ~~12~~ 15 16
18 ~~20~~ 21 27

20	6	
	12	

3 4 6 9 12
16 18 20 24

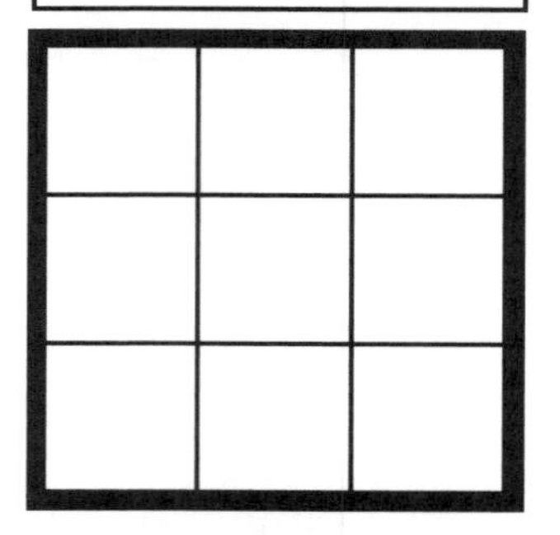

3 5 9 10 15
21 25 30 45

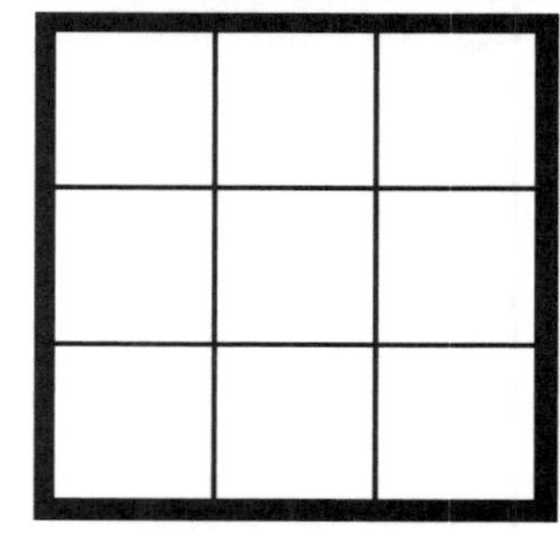

3 6 7 12 14
18 21 24 28

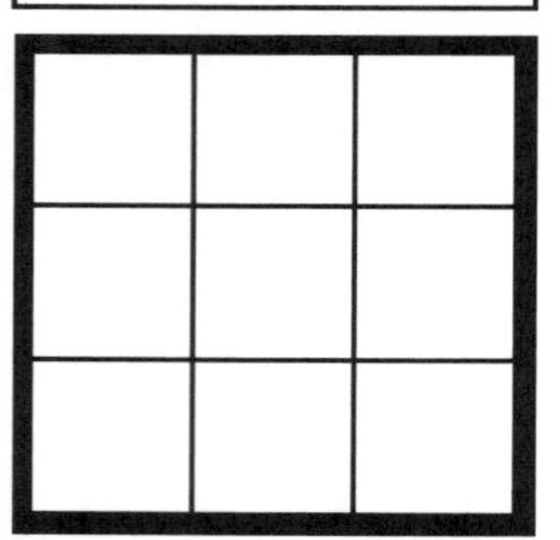

8 10 12 15 24
28 32 35 40

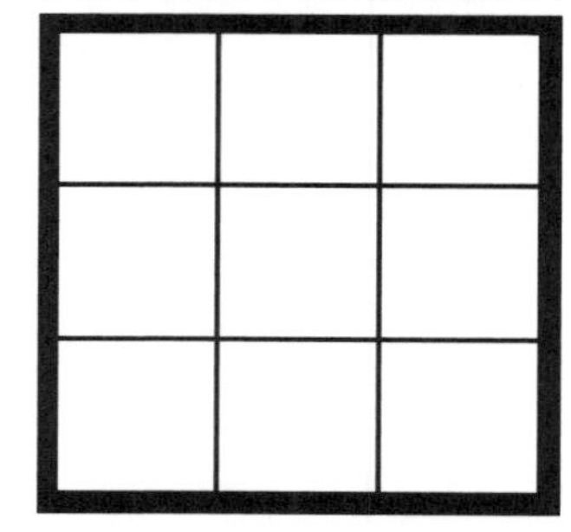

3 8 9 16 18
20 27 36 54

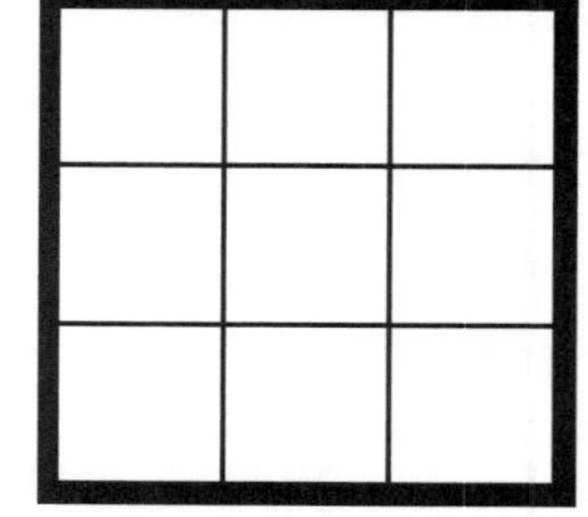

Date ____________ Name ____________________

Common Factors Practice Challenge

Carefully write the boxed numbers in each grid so that every number placed shares a common factor with every number in an adjacent cell horizontally, vertically, and diagonally.

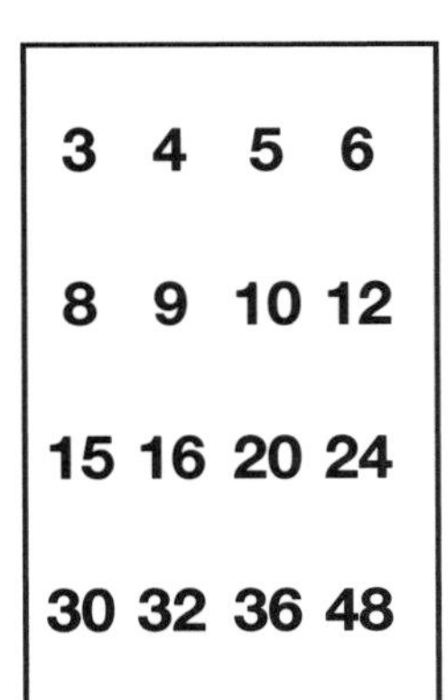

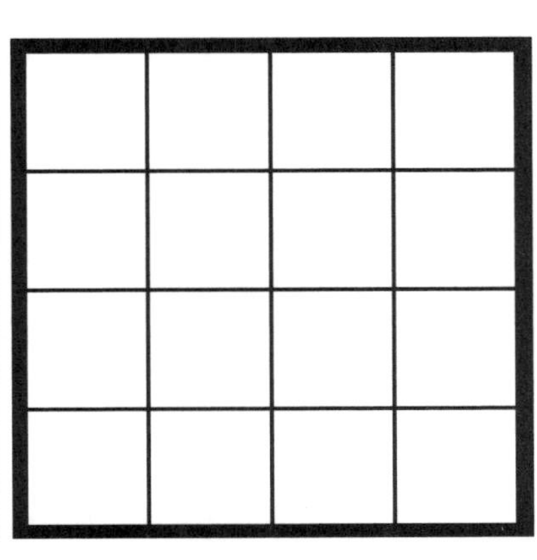

2 3 7 9
12 14 18 21
24 27 28 35
36 42 45 49

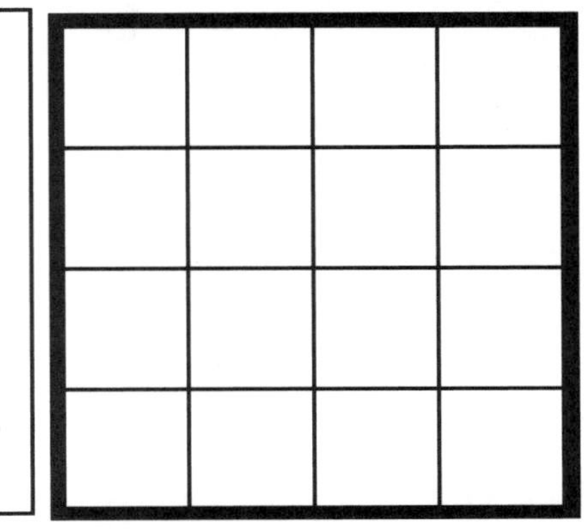

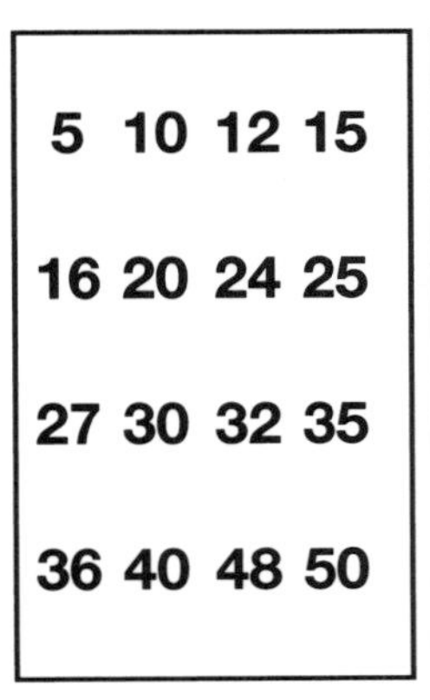

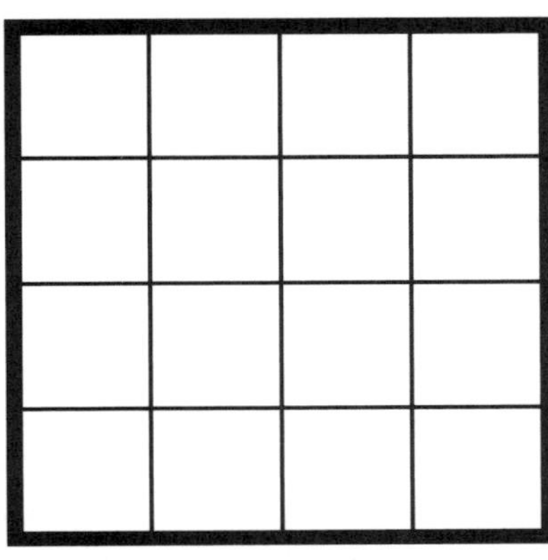

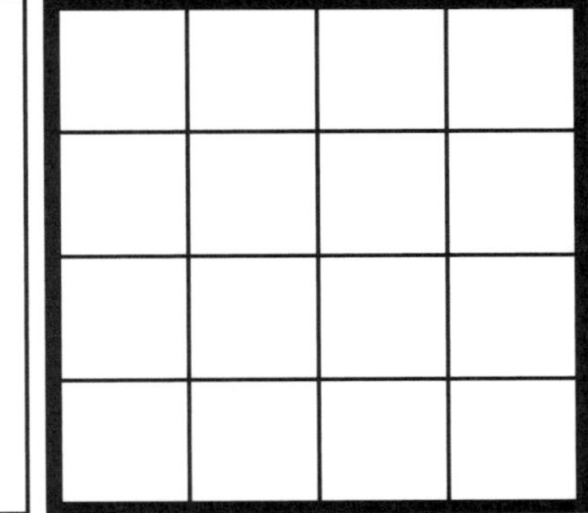

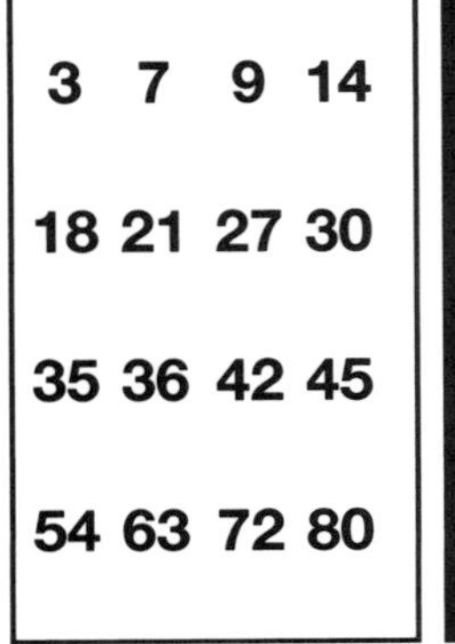

3 9 12 15
18 21 25 26
30 32 36 40
42 45 48 50
52 55 57 63

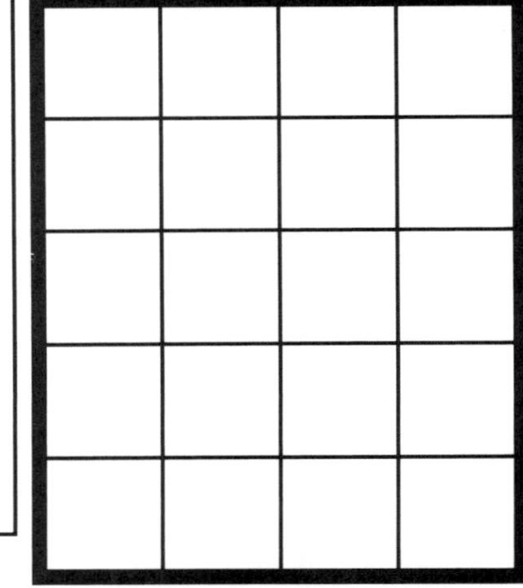

Date ____________ Name ____________________

Using LCM and GFC to Check Multiplication

Write the missing numbers. The first problem has been done for you.

1. 10 × 6 = 30 30 × 2 = 30 LCM GCF	**2.** 20 × 24 = ____ ____ × ____ = ____ LCM GCF	**3.** 12 × 15 = ____ ____ × ____ = ____ LCM GCF
4. 20 × 32 = ____ ____ × ____ = ____ LCM GCF	**5.** 36 × 28 = ____ ____ × ____ = ____ LCM GCF	**6.** 28 × 40 = ____ ____ × ____ = ____ LCM GCF
7. 36 × 45 = ____ ____ × ____ = ____ LCM GCF	**8.** 15 × 48 = ____ ____ × ____ = ____ LCM GCF	**9.** 45 × 27 = ____ ____ × ____ = ____ LCM GCF
10. 14 × 21 = ____ ____ × ____ = ____ LCM GCF	**11.** 36 × 15 = ____ ____ × ____ = ____ LCM GCF	**12.** 25 × 24 = ____ ____ × ____ = ____ LCM GCF
13. 50 × ____ = 1500 ____ × ____ = ____ LCM GCF	**14.** 15 × ____ = 750 ____ × ____ = ____ LCM GCF	**15.** 28 × ____ = 224 ____ × ____ = ____ LCM GCF

Think: What patterns do you notice?

Fractions

Assumptions Fraction operations have previously been taught and reviewed, emphasizing understanding and building fraction sense. Concrete objects and visual models, such as fraction circles or rectangles and grids, have been used extensively.

Section Overview and Suggestions

Sponges

Fitting Fractions pp. 48–49

Fraction Three-in-a-Row pp. 50–52

These open-ended, repeatable whole-class or small-group warm-ups reinforce all fraction operations and require mental computation. Frequent use of *Fitting Fractions* will ensure greater success with the Games and Independent Activities in this section.

Skill Checks

Partial Possibilities 1-6 pp. 53–55

These provide a way for parents, students, and you to see students' improvement with fraction computation. Copies can be cut in half so that each check may be used at a different time. Remember to have all students respond to STOP, the number sense task, before solving the ten problems.

Games

Fraction Formulation pp. 56–58

Seeking Fractions pp. 59–60

These open-ended and repeatable Games actively involve students in varied fraction operations as they create appropriate fractions and compute mentally. The first recording sheet of *Fraction Formulation* reinforces addition and subtraction while the second sheet reinforces multiplication and division.

Independent Activities

Fitting Fractions Practice pp. 61–62

Creating and Computing Fractions pp. 63–64

Each Independent Activity requires students to practice fraction operations to solve equations. Students mentally compute many additional problems as they seek equations that work. Both activities provide two versions, allowing practice of addition and subtraction or multiplication and division. The simple formats can be easily varied with new digits to provide long-term fraction computation practice with any or all operations.

Fitting Fractions

Tip The leader can draw an additional Digit Square and let students discard one drawn number in a "reject box."

Topic: Computing with Fractions

Object: Create fraction equations or inequalities that meet given criteria.

Groups: Whole class or small group

Materials

- set of Digit Squares (0 removed) for each student, p. 147
- transparency of *Fitting Fractions* with forms cut apart, p. 49
- set of transparent Digit Squares in opaque container

Directions

1. The leader describes a desired outcome for an equation or inequality.

   Example: Create an addition inequality showing a sum that is less than 1.
2. The leader displays the inequality format (see illustration) which is copied by the students. The leader explains that four Digit Squares will be drawn and announced one at a time.

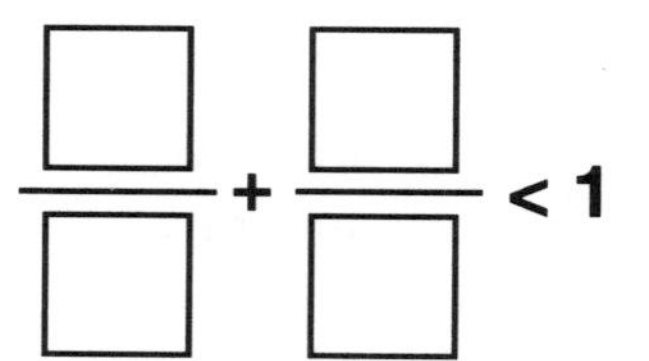

3. The leader draws the first Digit Square and announces the digit. The leader makes sure that each student places the announced digit in one of the squares before the leader draws the next Digit Square.
4. Digit Squares are not returned to the container until the end of the round. Students may not change or move placed Digit Squares.
5. This procedure is followed until four Digit Squares are drawn.
6. After students place the four Digit Squares, they carry out the computation to determine whether they have met the desired outcome. The leader might require students to prove they have made a true equation or inequality.
7. After students play additional rounds with the same outcome, the outcome can be changed.

   Possible outcomes for future rounds: sum closest to 1, difference closest to $\frac{1}{2}$, product less than $\frac{1}{4}$, quotient greater than 2
8. Vary this activity and increase fraction sense by announcing all four digits before students place their Digit Squares.

   Example: Use 2, 3, 5, and 6 to create the least possible difference.

Making Connections

Promote reflection and make mathematical connections by asking:

- How close was your answer to meeting the criteria?
- How did you decide where to place certain digits?

Fitting Fractions

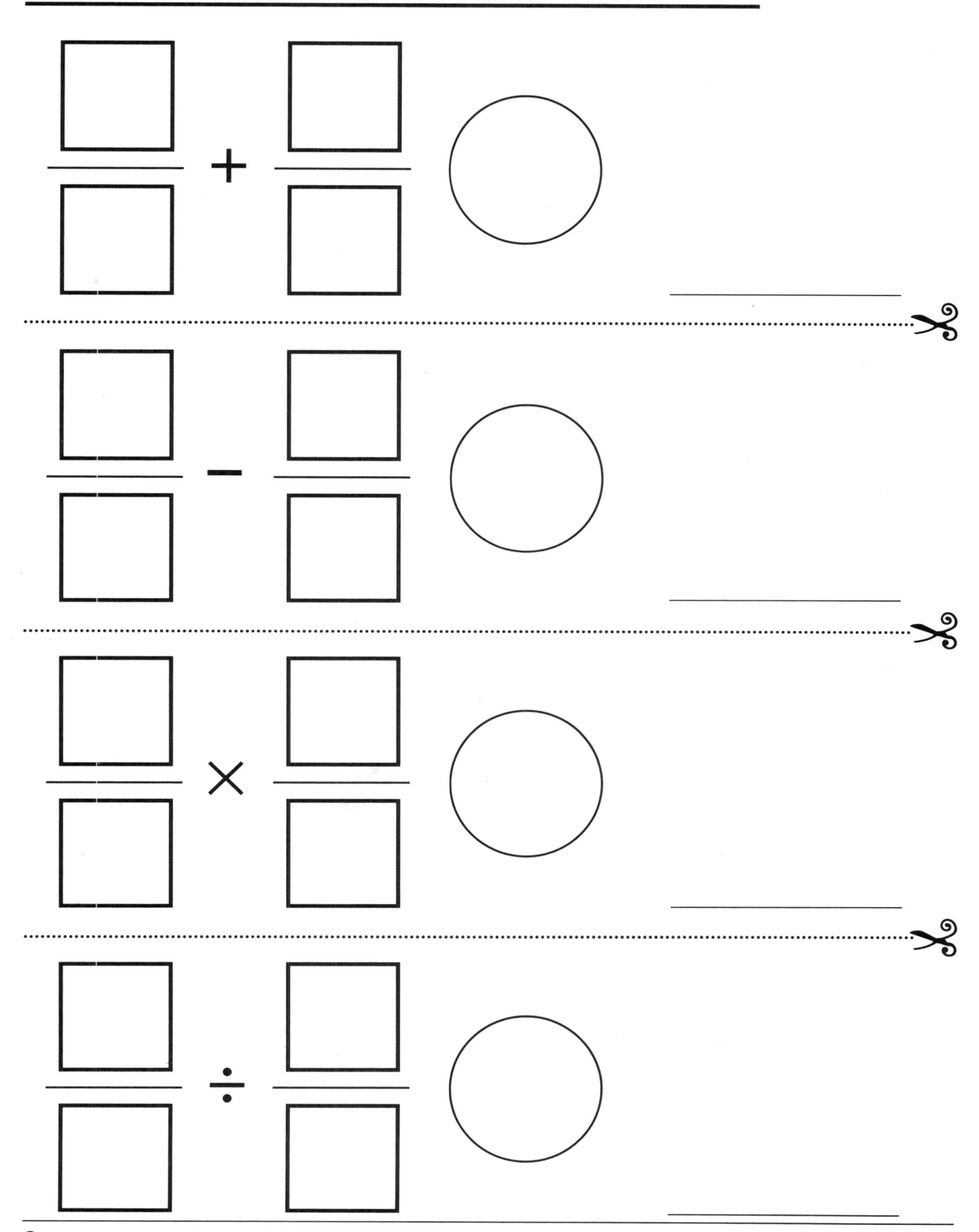

Fraction Three-in-a-Row

Topic: Fraction Sense and Computing with Fractions

Object: Cover three cells in a row.

Groups: Whole class or small group

Tip *If students find the level of clues difficult, have students pair up to solve the clues, or provide easier clues.*

Materials

- *Fraction Three-in-a-Row* form for each student, p. 51
- transparency of prepared clues, cut apart and placed in a container, p. 52
- 9 markers for each student
- scratch paper

Directions

1. Students create a unique gameboard by randomly recording each of the following fractions on their *Fraction Three-in-a-Row* form: $\frac{1}{3}, \frac{2}{5}, \frac{2}{3}, \frac{3}{7}, \frac{7}{8}, 1\frac{1}{3}, \frac{7}{2}, 3\frac{1}{5}, \frac{15}{7}$
2. The leader draws, reads, and displays a Round 1 Clue from the container.

   Example: "Thirteen tenths plus eleven fifths"
3. Students individually solve the posed problem and cover the corresponding fraction with a marker.
4. After students have responded, the leader continues to follow this procedure of providing clues while students cover the corresponding fractions.
5. When students have three markers in a row, they call out, "Three in a row." The activity can continue until all students have at least one "three in a row." After students share their three-in-a-row fractions, some might challenge the results and request proof.
6. The leader uses the Round 2 Clues to provide a repeat round of this activity with the same playing board.
7. Next, students collaborate to provide nine additional fraction clues for future rounds.
8. The leader or students can identify nine "new" fractions for future playing boards and nine corresponding clues.

Making Connections

Promote reflection and make mathematical connections by asking:

- Which clues were you able to figure out mentally? Please explain.
- What was helpful in creating clues for future rounds?

Fraction Three-in-a-Row

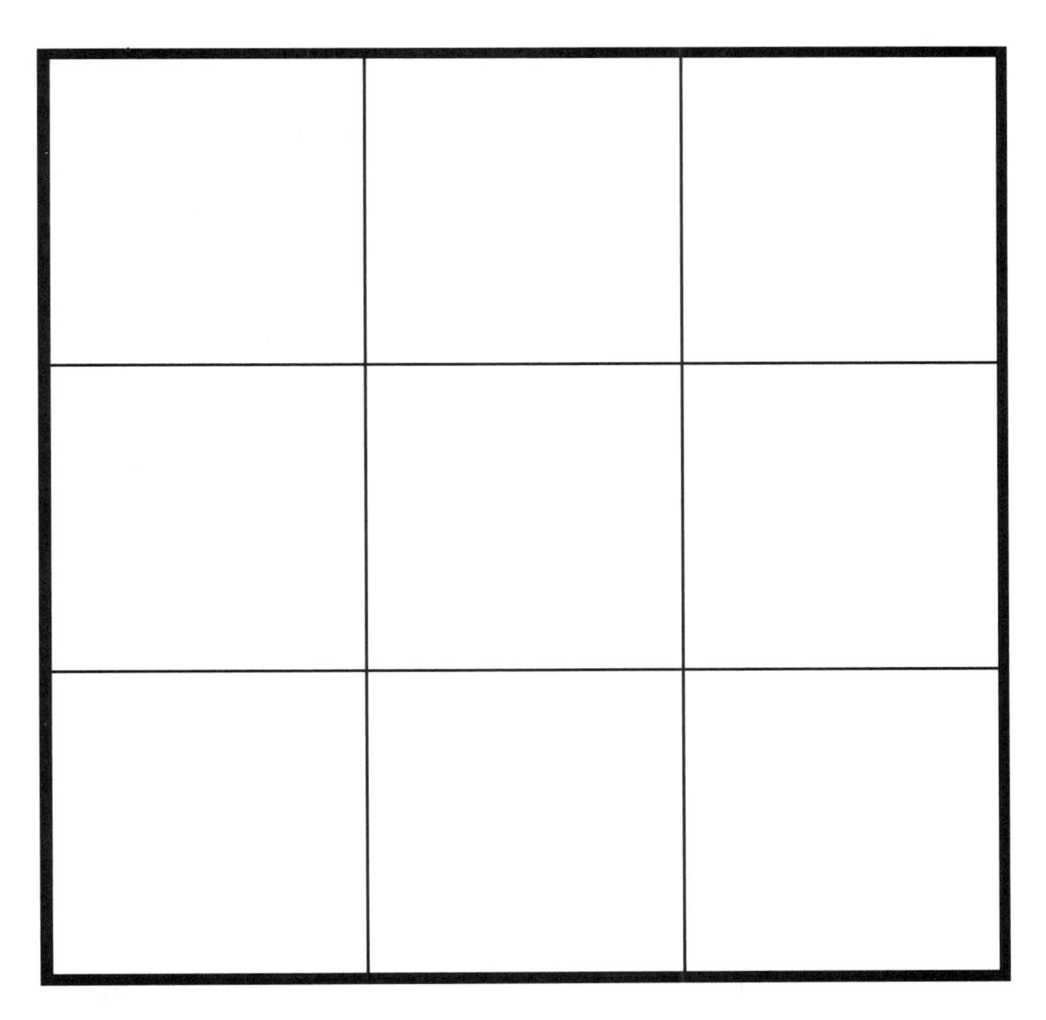

Fraction Three-in-a-Row Clues

Gameboard Fractions	Round 1 Clues	Round 2 Clues
$\frac{1}{3}$	$\frac{1}{2}$ of $\frac{2}{3}$	$1\frac{1}{2} \div 4\frac{1}{2}$
$\frac{2}{5}$	$\frac{3}{5}$ of $\frac{2}{3}$	$1\frac{1}{4} - \frac{17}{20}$
$\frac{2}{3}$	$\frac{1}{2} + \frac{1}{6}$	$\frac{2}{9} \div \frac{1}{3}$
$\frac{3}{7}$	reciprocal of $2\frac{1}{3}$	$1\frac{1}{7} - \frac{15}{21}$
$\frac{7}{8}$	$2\frac{3}{8} - 1\frac{1}{2}$	$\frac{5}{8} \div \frac{5}{7}$
$1\frac{1}{3}$	$\frac{14}{15} + \frac{3}{5} - \frac{2}{10}$	$\frac{2}{3} \div \frac{1}{2}$
$\frac{7}{2}$	$\frac{13}{10} + \frac{11}{5}$	$3 \times \frac{7}{6}$
$3\frac{1}{5}$	$12 \div 3\frac{3}{4}$	$8 \div 2\frac{1}{2}$
$\frac{15}{7}$	rounds to 2	$1\frac{5}{7} + \frac{9}{21}$

Date ____________ Name ____________

Partial Possibilities 1

STOP Don't start yet! Star two problems that may have answers less than $\frac{1}{2}$.
Express all fractions in lowest terms.

1. $\frac{1}{4} + \frac{2}{3}$

2. $\frac{3}{5} + \frac{3}{4}$

3. $\frac{3}{4} - \frac{1}{2}$

4. $3 - \frac{2}{7}$

5. $1\frac{1}{3} + \frac{2}{9} =$ ______

6. $\frac{5}{8} - \frac{1}{3} =$ ______

7. $4\frac{1}{5} - 2\frac{1}{2} =$ ______

8. $\frac{3}{4} \times \frac{1}{2} =$ ______

9. $\frac{2}{3} \times 1\frac{4}{5} =$ ______

10. $\frac{3}{5} \div \frac{1}{3} =$ ______

Go On What comes next? $3\frac{2}{3}$, $3\frac{4}{9}$, $3\frac{2}{9}$, ______, ______, ______
Describe the pattern.

Date ____________ Name ____________

Partial Possibilities 2

STOP Don't start yet! Star two problems that may have answers greater than 2.
Express all fractions in lowest terms.

1. $\frac{2}{3} + \frac{1}{6}$

2. $\frac{5}{6} + \frac{5}{8}$

3. $\frac{7}{8} - \frac{1}{4}$

4. $4 - \frac{5}{9}$

5. $1\frac{3}{4} + \frac{5}{12} =$ ______

6. $\frac{5}{6} - \frac{1}{4} =$ ______

7. $3\frac{2}{3} - 1\frac{4}{5} =$ ______

8. $\frac{3}{8} \times \frac{2}{3} =$ ______

9. $\frac{3}{7} \times 2\frac{5}{8} =$ ______

10. $\frac{2}{3} \div \frac{3}{7} =$ ______

Go On Name three different fractions close to but not equal to $\frac{3}{4}$.

Date ____________ Name ____________

Partial Possibilities 3

Don't start yet! Star two problems that may have answers close to 1. Express all fractions in lowest terms.

1. $\begin{array}{r} \frac{3}{4} \\ + \frac{1}{8} \\ \hline \end{array}$

2. $\begin{array}{r} \frac{2}{3} \\ + \frac{5}{7} \\ \hline \end{array}$

3. $\begin{array}{r} \frac{1}{2} \\ - \frac{1}{3} \\ \hline \end{array}$

4. $\begin{array}{r} 3 \\ - \frac{3}{5} \\ \hline \end{array}$

5. $2\frac{2}{5} + \frac{3}{10} =$ ______

6. $\frac{2}{3} - \frac{4}{7} =$ ______

7. $4\frac{1}{4} - 2\frac{2}{3} =$ ______

8. $\frac{5}{6} \times \frac{1}{2} =$ ______

9. $\frac{3}{5} \times 1\frac{5}{6} =$ ______

10. $\frac{3}{7} \div \frac{2}{5} =$ ______

What comes next? $2\frac{7}{12}$, $2\frac{5}{12}$, $2\frac{1}{4}$, ______, ______, ______
Describe the pattern.

Date ____________ Name ____________

Partial Possibilities 4

Don't start yet! Star two problems that may have answers between $\frac{1}{2}$ and 1. Express all fractions in lowest terms.

1. $\begin{array}{r} \frac{5}{12} \\ + \frac{1}{4} \\ \hline \end{array}$

2. $\begin{array}{r} \frac{4}{5} \\ + \frac{2}{3} \\ \hline \end{array}$

3. $\begin{array}{r} \frac{5}{6} \\ - \frac{1}{2} \\ \hline \end{array}$

4. $\begin{array}{r} 4 \\ - \frac{7}{9} \\ \hline \end{array}$

5. $1\frac{1}{8} + \frac{3}{4} =$ ______

6. $\frac{4}{5} - \frac{1}{3} =$ ______

7. $3\frac{5}{12} - 1\frac{5}{6} =$ ______

8. $\frac{3}{8} \times \frac{4}{7} =$ ______

9. $\frac{5}{7} \times 1\frac{3}{4} =$ ______

10. $\frac{5}{6} \div \frac{3}{7} =$ ______

Go On Name three different fractions between $\frac{1}{4}$ and $\frac{1}{2}$.

Date ______________ Name ______________________

Partial Possibilities 5

Don't start yet! Star two problems that may have answers between 1 and 2. Express all fractions in lowest terms.

1. $\frac{3}{5} + \frac{1}{4}$

2. $\frac{5}{8} + \frac{2}{3}$

3. $\frac{3}{4} - \frac{1}{3}$

4. $3 - \frac{5}{7}$

5. $2\frac{1}{6} + \frac{1}{3} =$ ______

6. $\frac{4}{5} - \frac{1}{3} =$ ______

7. $3\frac{3}{8} - 1\frac{1}{2} =$ ______

8. $\frac{5}{8} \times \frac{3}{4} =$ ______

9. $\frac{4}{5} \times 2\frac{1}{7} =$ ______

10. $\frac{7}{8} \div \frac{2}{7} =$ ______

Go On What comes next? $2\frac{7}{8}$, $2\frac{11}{24}$, $2\frac{1}{24}$, ______, ______, ______
Describe the pattern.

Date ______________ Name ______________________

Partial Possibilities 6

Don't start yet! Star two problems that may have answers less than $\frac{1}{2}$. Express all fractions in lowest terms.

1. $\frac{2}{3} + \frac{1}{8}$

2. $\frac{3}{4} + \frac{5}{6}$

3. $\frac{2}{3} - \frac{5}{12}$

4. $4 - \frac{7}{8}$

5. $1\frac{3}{5} + \frac{3}{10} =$ ______

6. $\frac{6}{7} - \frac{3}{5} =$ ______

7. $4\frac{3}{5} - 2\frac{2}{3} =$ ______

8. $\frac{2}{3} \times \frac{4}{9} =$ ______

9. $\frac{5}{6} \times 1\frac{7}{10} =$ ______

10. $\frac{4}{9} \div \frac{5}{6} =$ ______

Go On Name two different fractions between 0 and $\frac{1}{12}$ whose numerators are not 1.

Fraction Formulation

Topic: Mentally Computing with Fractions

Object: Create fraction inequalities with the greatest sum and least difference.

Groups: Pair players

Materials for each group

- *Fraction Formulation A* recording sheet, p. 57
- Digit Squares (1–6 only) in opaque container, p. 147

Tips: If students become confident playing this version, extend the game to use the digits 1 through 9. Use Fraction Formulation B, *p. 58, to practice multiplication and division of fractions.*

Directions

1. Each pair draws two Digit Squares from their separate containers. The pairs discuss and decide where to place each of the drawn digits.
2. After each pair records their placement of the Digit Squares, they show the other pair and return the Digit Squares to their container. A pair may not create more than two improper fractions during any round.
3. This procedure continues until both pairs have completed both inequalities by filling in twelve digits. A pair can decide not to accept and record one or both of their drawn digits.
4. The pair that accurately completes the two inequalities in the fewest turns receives 1 point. It's possible that both pairs might qualify.
5. After both pairs verify their final results, the pairs find the exact sum and difference for their two inequalities. The pair with the greater sum receives 1 point and the pair with the smaller difference receives 1 point.
6. The pairs play an additional round and determine which pair has accumulated the most points.

Pair A-6 turns

$$\underbrace{\frac{3}{2} + \frac{6}{1}}_{7\frac{1}{2}} > \frac{1}{5}$$

$$\underbrace{\frac{4}{6} - \frac{1}{2}}_{\frac{1}{6}} < \frac{3}{5}$$

Pair B-6 turns

$$\underbrace{\frac{3}{1} + \frac{5}{2}}_{5\frac{1}{2}} > \frac{1}{6}$$

$$\underbrace{\frac{2}{6} - \frac{1}{3}}_{0} < \frac{4}{5}$$

Points	Team A	Team B
Fewest Turns	1	1
Larger Sum	1	0
Smaller Difference	0	1
TOTALS	2	2

Making Connections

Promote reflection and make mathematical connections by asking:

- How would you play differently in future games?
- Where would you recommend placing 2s . . . 6s?

Fraction Formulation A

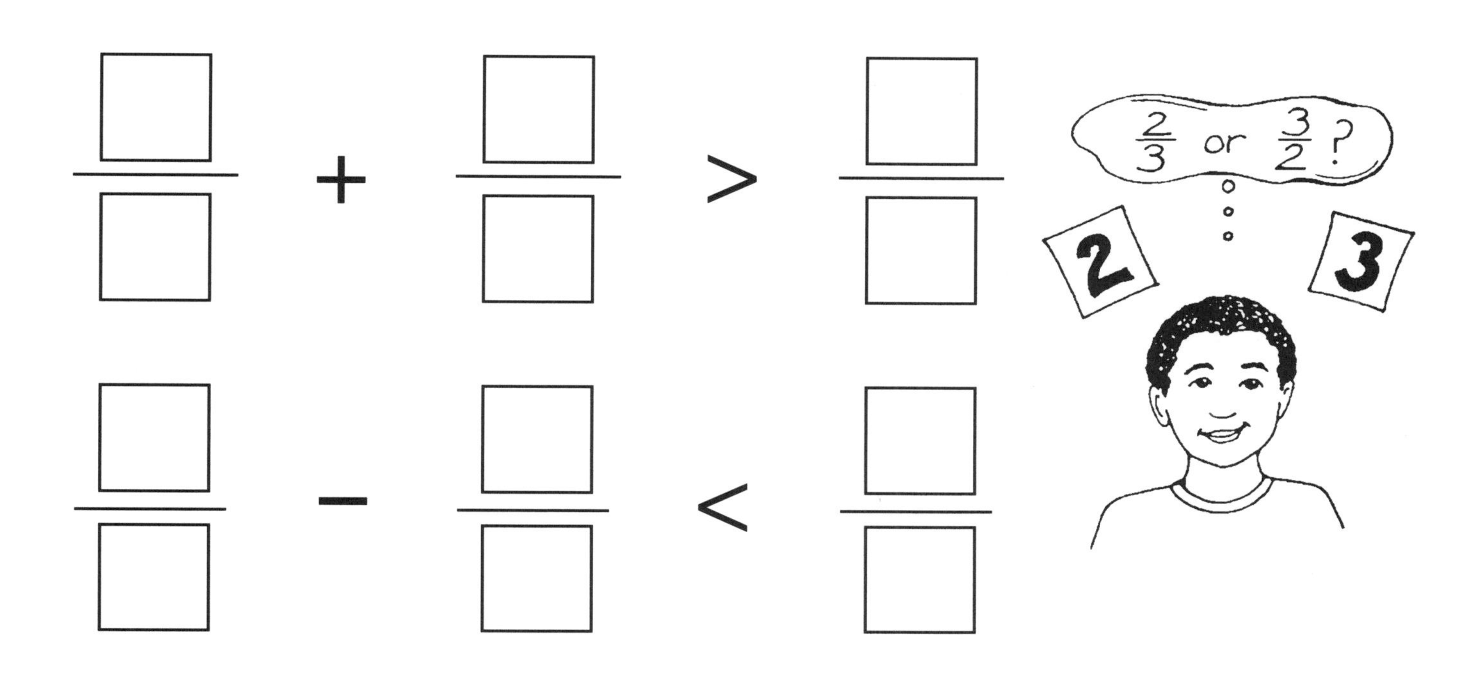

Fraction Formulation A

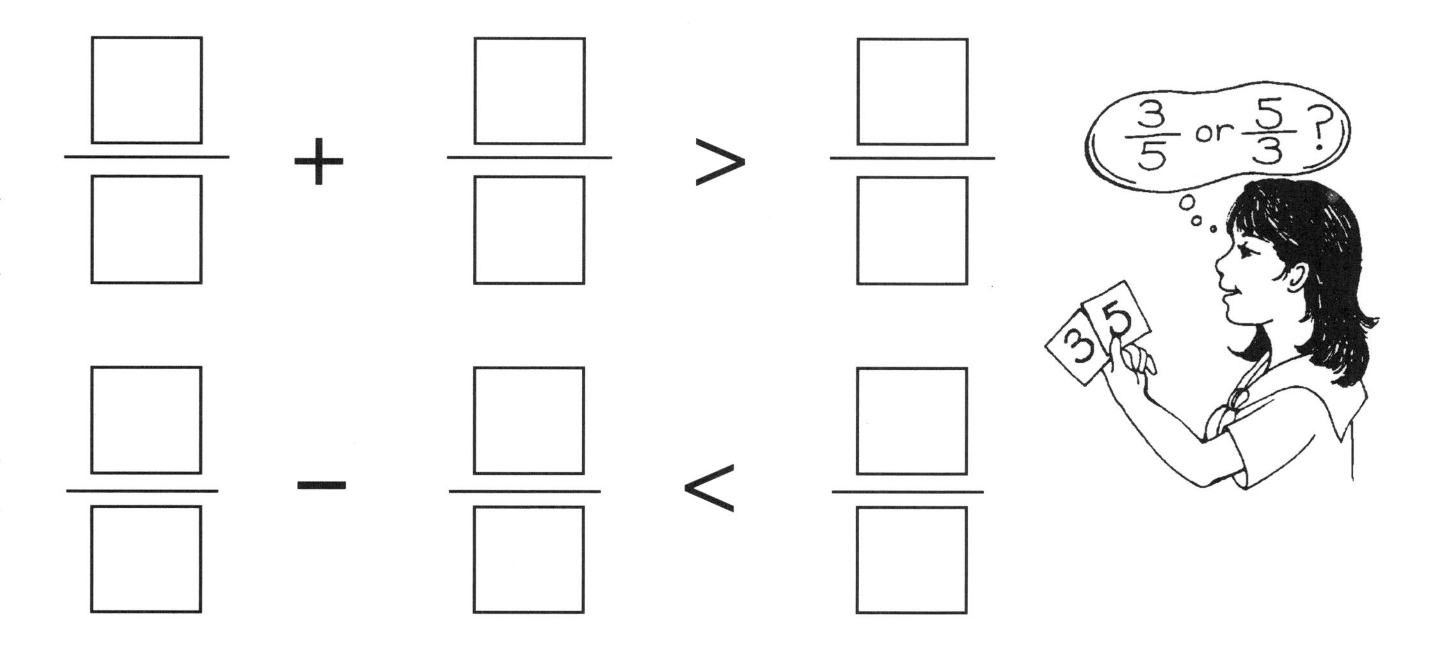

Fraction Formulation B

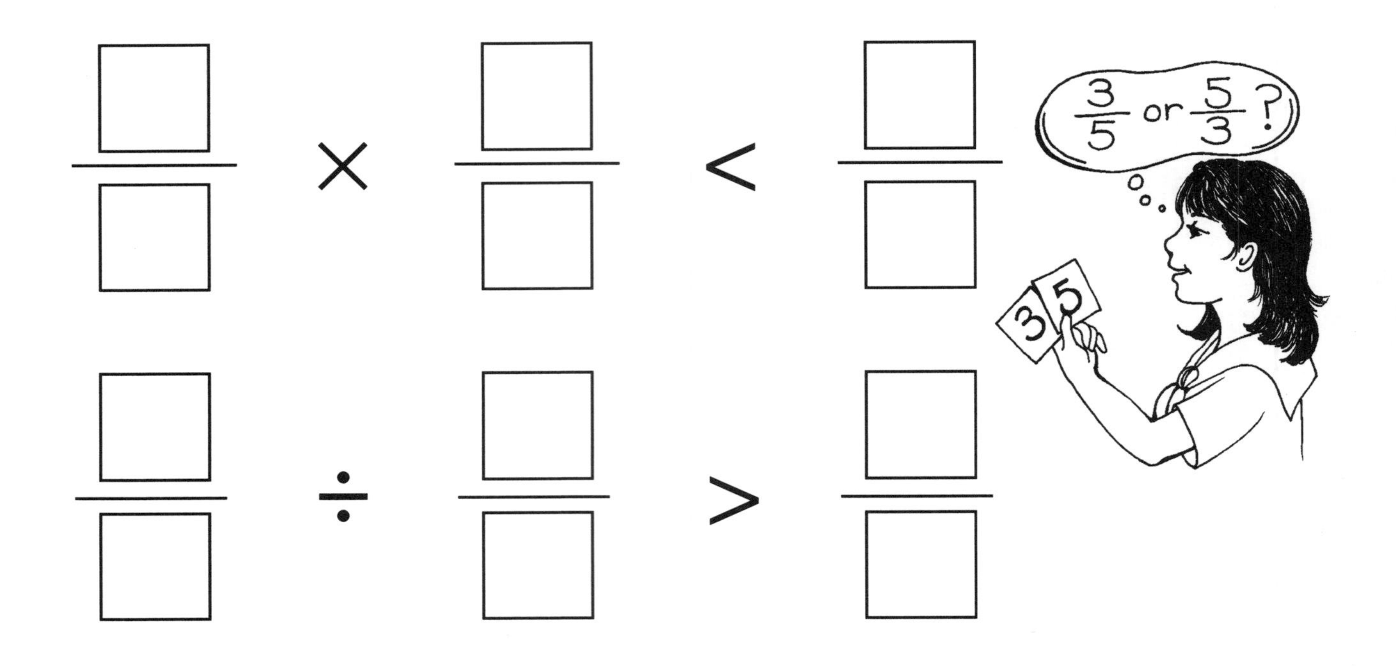

Fraction Formulation B

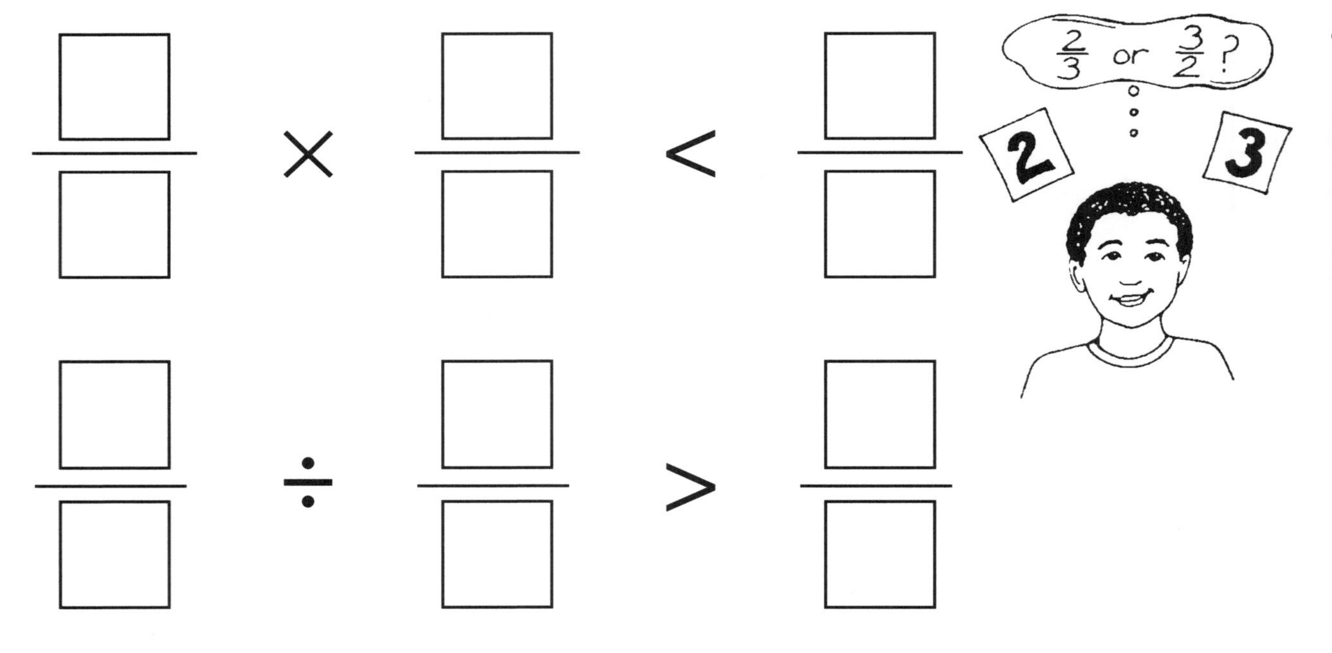

Seeking Fractions

Topic: Computing with Fractions Mentally

Object: Create fraction equations and inequalities with mixed operations.

Groups: 2–4 players

Materials for each group

- *Seeking Fractions* recording sheet for each player, p. 60
- 4 Number Cubes or 4 sets of Digit Squares (1–6 only), p. 147
- scratch paper

Tips *To increase possibilities and success rate, have players roll five number cubes and select four of the five numbers to create expressions. Some students benefit by writing the rolled numbers on small movable squares of paper that can be easily rearranged.*

Directions

1. One player rolls the four Number Cubes. (If preferred, four Digit Squares can be drawn and substituted for Number Cubes. Digit Squares should be replaced at the end of each turn.)
2. All players use the same four rolled numbers to create and record a fraction equation or inequality that satisfies one of the listed conditions.

   Example: 2, 3, 4, and 6 are rolled. One player records $\frac{2}{3} + \frac{4}{6} = 1\frac{1}{3}$ (mixed number) while another player records $\frac{3}{6} \times \frac{4}{2} = 1$ (whole number).

I ☑ added ☑ subtracted ☑ multiplied ☐ divided	
$\frac{\square}{\square} \bigcirc \frac{\square}{\square} = \frac{1}{2}$	$\frac{6}{3} \otimes \frac{4}{2} = 1$ (whole number)
$\frac{6}{3} \ominus \frac{4}{2} < 1$	$\frac{\square}{\square} \bigcirc \frac{\square}{\square} > \frac{1}{3}$ and < 1
$\frac{\square}{\square} \bigcirc \frac{\square}{\square} > 1$	$\frac{2}{3} \oplus \frac{4}{6} > 1\frac{1}{3}$ (mixed number)

3. When all players have recorded their equations or inequalities, they check off any operation that was used and share the results.
4. Players repeat the process using four more rolled or drawn numbers.
5. Each player is required to use all four operations among the equations and inequalities. If a player is unable to create a fraction equation or inequality that matches a description, the player records nothing for that turn.
6. The winner is the first player to complete six qualifying fraction equations and inequalities using all operations.
7. Since students develop winning strategies as they play, it is recommended that additional rounds be played. Players will need additional *Seeking Fractions* recording sheets.

Making Connections

Promote reflection and make mathematical connections by asking:

- Which equations were easier to create? Why?
- What strategy helped you record on almost every turn?

Seeking Fractions

Recording Sheet

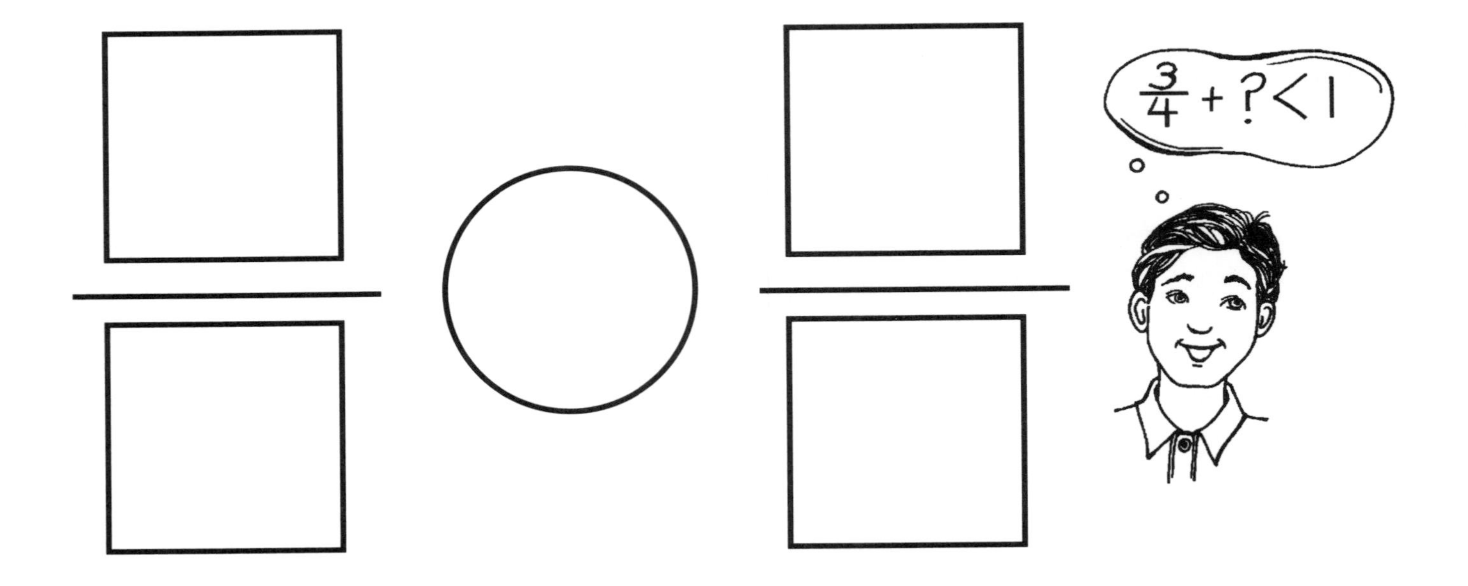

I added subtracted multiplied divided

= $\frac{1}{2}$

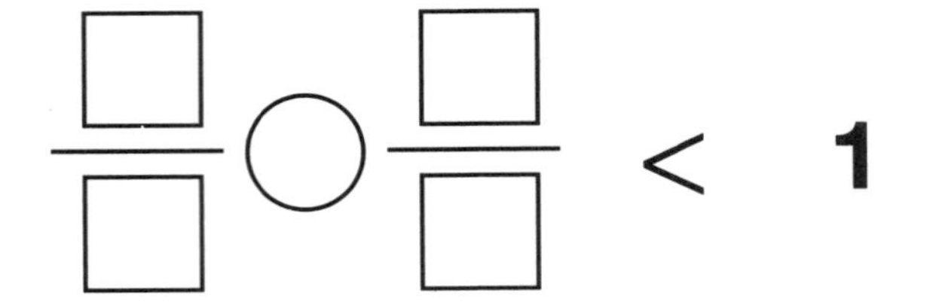

> 1

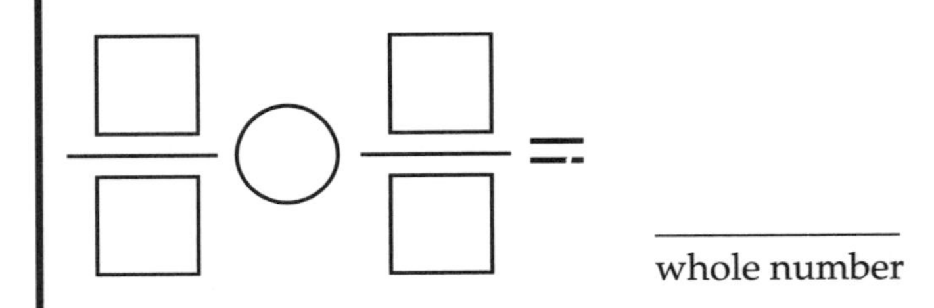

> $\frac{1}{3}$ and < 1

> mixed number

Date ____________ Name ____________

Fitting Fractions Practice I

1. Use 1, 2, 3, and 6 to get the greatest sum.

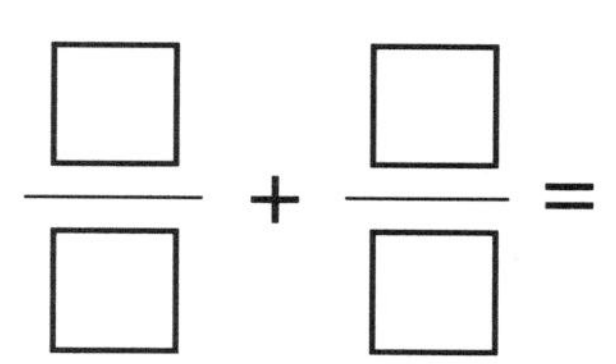

2. Use 1, 2, 3, and 6 to get the smallest sum.

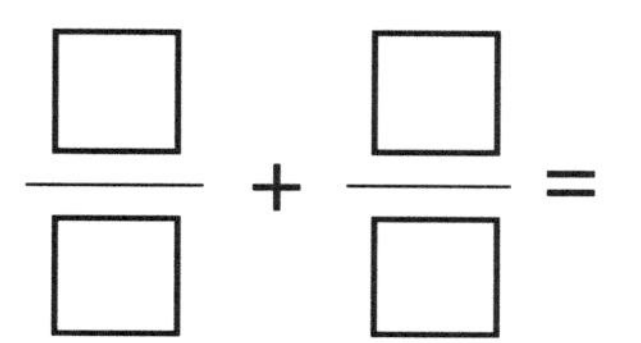

3. Use 1, 2, 3, and 4 to get the smallest difference.

4. Use 1, 2, 3, and 6 to get the greatest difference.

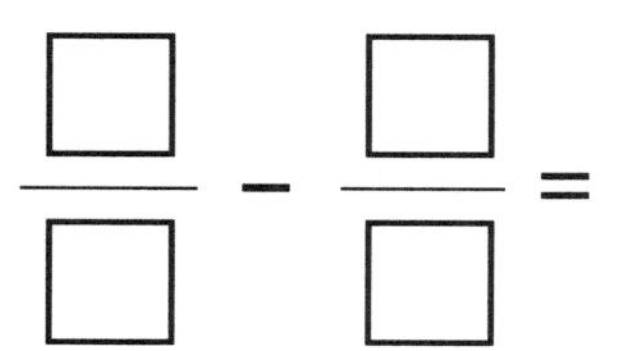

5. Use 1, 2, 3, and 6 to get a difference of $\frac{1}{2}$.

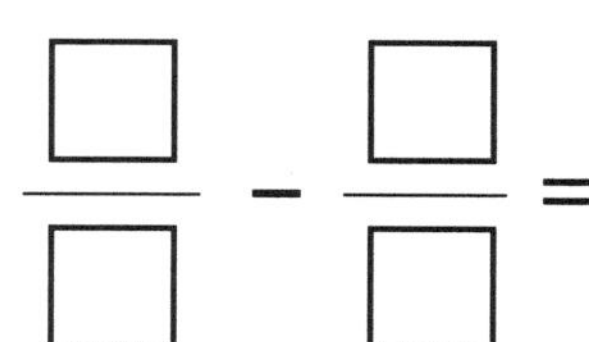

6. Use 1, 2, 3, and 4 to get the smallest sum.

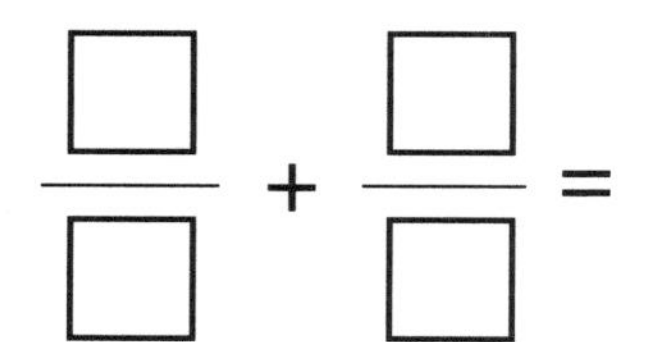

7. Use 1, 2, 3, and 4 to get a sum greater than 2.

Date ____________________ Name ________________________________

Fitting Fractions Practice II

1. Use 1, 2, 3, and 6 to get the greatest product.

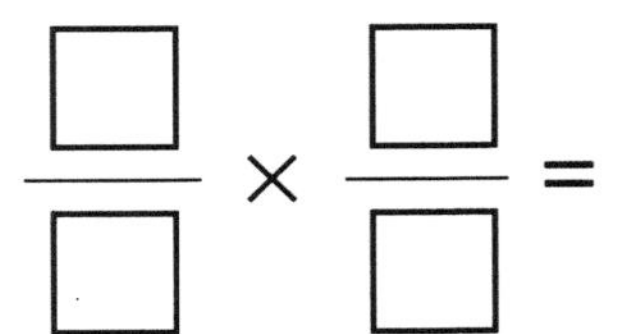

2. Use 1, 2, 3, and 6 to get the smallest product.

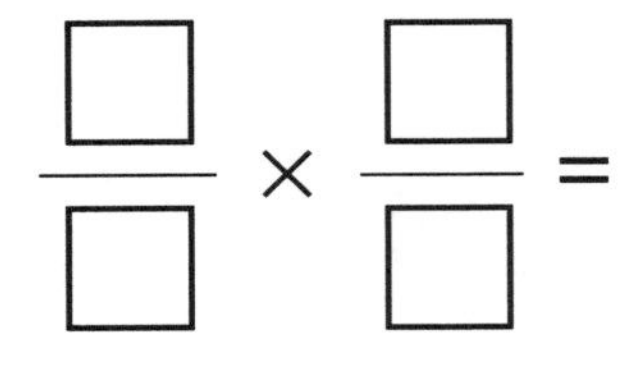

3. Use 1, 2, 3, and 4 to get the smallest quotient.

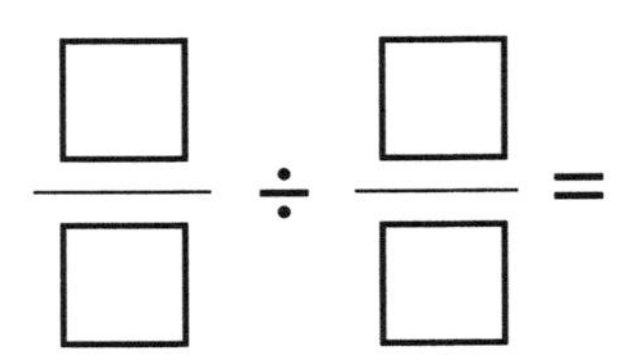

4. Use 1, 2, 3, and 6 to get the greatest quotient.

$\frac{\square}{\square} \div \frac{\square}{\square} =$

5. Use 1, 2, 3, and 4 to get a product less than $\frac{1}{2}$.

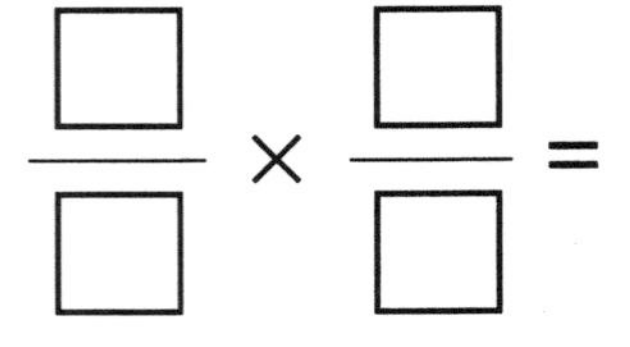

6. Use 1, 2, 3, and 4 to get the largest product.

$\frac{\square}{\square} \times \frac{\square}{\square} =$

7. Use 2, 3, 4, and 5 to get a product greater than 2.

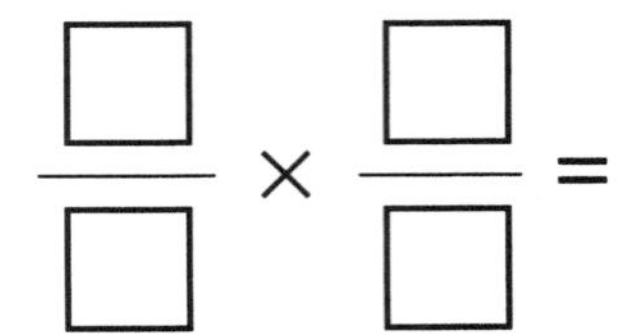

$\frac{4}{3}$? $\frac{1}{6}$? $\frac{3}{6}$? $\frac{6}{4}$?

Date ______________ Name ____________________

Creating and Computing Fractions I

For each box in each equation or inequality, use any of these numbers:

1 2 3 4 8 12 16

Do not use a number more than once in each sentence. Avoid using improper fractions.

1. $\frac{\square}{\square} + \frac{\square}{\square} > \frac{\square}{\square}$

2. $\frac{\square}{\square} + \frac{\square}{\square} < \frac{\square}{\square}$

3. $\frac{\square}{\square} - \frac{\square}{\square} < \frac{\square}{\square}$

4. $\frac{\square}{\square} - \frac{\square}{\square} = \frac{1}{2}$

For each box in each equation or inequality, use any of these numbers:

1 2 3 5 6 10 15

Do not use a number more than once in each sentence. Avoid using improper fractions.

5. $\frac{\square}{\square} + \frac{\square}{\square} > \frac{\square}{\square}$

6. $\frac{\square}{\square} + \frac{\square}{\square} = \frac{3}{5}$

7. $\frac{\square}{\square} - \frac{\square}{\square} < \frac{\square}{\square}$

8. $\frac{\square}{\square} - \frac{\square}{\square} > \frac{\square}{\square}$

For each box in each equation or inequality, use any of these numbers:

1 2 3 6 8 9 12

Do not use a number more than once in each sentence. Avoid using improper fractions.

9. $\frac{\square}{\square} + \frac{\square}{\square} > \frac{\square}{\square}$

10. $\frac{\square}{\square} + \frac{\square}{\square} < \frac{\square}{\square}$

11. $\frac{\square}{\square} - \frac{\square}{\square} < \frac{\square}{\square}$

12. $\frac{\square}{\square} - \frac{\square}{\square} = \frac{2}{3}$

Date ____________ Name ____________

Creating and Computing Fractions II

For each box in each equation or inequality, use any of these numbers:

1 2 3 4 5 7 8

Do not use a number more than once in each sentence. Avoid using improper fractions.

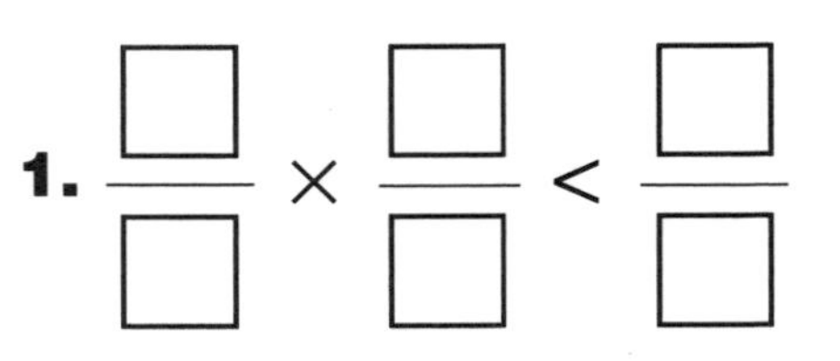

1. $\frac{\square}{\square} \times \frac{\square}{\square} < \frac{\square}{\square}$

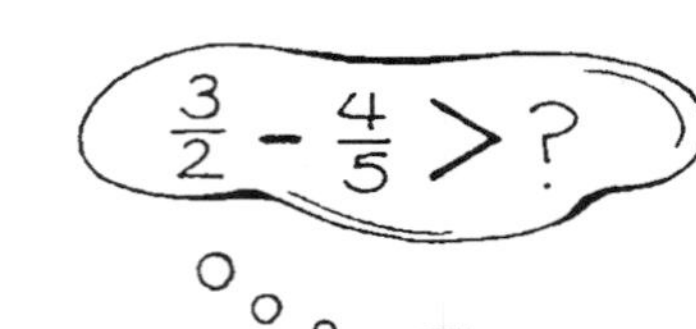

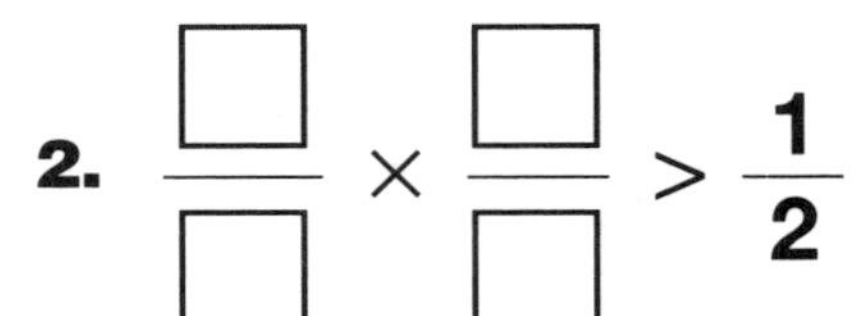

2. $\frac{\square}{\square} \times \frac{\square}{\square} > \frac{1}{2}$

3. $\frac{\square}{\square} \div \frac{\square}{\square} > \frac{\square}{\square}$

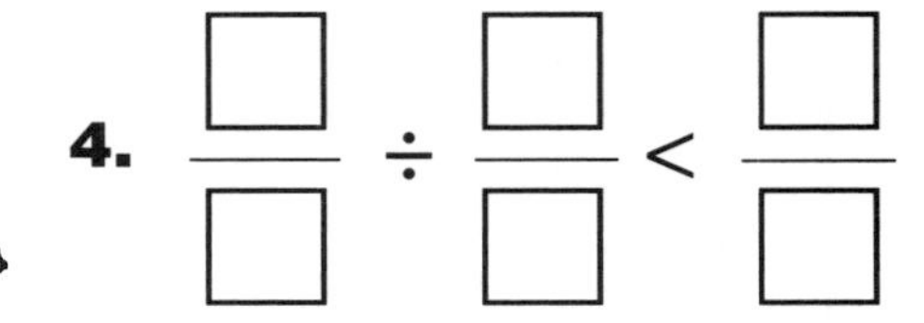

4. $\frac{\square}{\square} \div \frac{\square}{\square} < \frac{\square}{\square}$

For each box in each equation or inequality, use any of these numbers:

1 3 5 6 7 8 9

Do not use a number more than once in each sentence. Avoid using improper fractions.

5. $\frac{\square}{\square} \times \frac{\square}{\square} < \frac{\square}{\square}$

6. $\frac{\square}{\square} \times \frac{\square}{\square} > \frac{\square}{\square}$

7. $\frac{\square}{\square} \div \frac{\square}{\square} > \frac{\square}{\square}$

8. $\frac{\square}{\square} \div \frac{\square}{\square} < \frac{1}{2}$

For each box in each equation or inequality, use any of these numbers:

1 2 4 5 6 9 10

Do not use a number more than once in each sentence. Avoid using improper fractions.

9. $\frac{\square}{\square} \times \frac{\square}{\square} < \frac{\square}{\square}$

10. $\frac{\square}{\square} \times \frac{\square}{\square} = \frac{9}{25}$

11. $\frac{\square}{\square} \div \frac{\square}{\square} > \frac{\square}{\square}$

12. $\frac{\square}{\square} \div \frac{\square}{\square} < \frac{\square}{\square}$

Decimals

Assumptions Decimal operations have previously been taught and reviewed, emphasizing understanding and building decimal sense. Concrete objects and visual models, such as money or base-ten materials and grids, have been used extensively.

Section Overview and Suggestions

Sponges

Decimal Draw pp. 66–67

Mystery Factor pp. 68–69

Decimal Tic-Tac-Toe pp. 70–71

These repeatable whole-class or small-group warm-ups build decimal and number sense as students enhance their abilities to mentally compute with decimals.

Skill Checks

Partial Possibilities 7-12 pp. 72–74

These provide a way for parents, students, and you to see students' improvement with decimal conversion and computation. Copies can be cut in half so that each Check may be used at a different time. Remember to have all students respond to STOP, the number sense task, before solving the ten problems.

Games

Decimal Four-in-a-Row pp. 75–76

Decimal Bingo pp. 77–79

These open-ended and repeatable Games actively involve students in multiplying and dividing decimals, while students enhance their strategic thinking abilities. The two gameboards for *Decimal Bingo* allow long-term and home use of this challenging yet popular game.

Independent Activities

Decimal Trails pp. 80–81

Fitting Decimals pp. 82–83

Magic Squares p. 84

Each Independent Activity easily engages students to practice decimal operations. These challenging activities require much mental computation as students improve their number sense. Both *Decimal Trails* and *Fitting Decimals* include two versions, allowing practice of addition and subtraction or multiplication and division. Further reasoning will be enhanced if students are encouraged to create similar activities for their classmates to solve.

Decimal Draw

Topic: Adding and Subtracting Decimals

Object: Compute an answer as close as possible to the target number.

Groups: Whole class or small group

Tips It's important that every student record each drawn digit before the next digit is drawn. Consider decimal target numbers for future rounds.

Materials

- *Decimal Draw* recording sheet for each student, p. 67
- 2 sets of transparent Digit Squares in opaque container, p. 147

Directions

1. The leader announces a target number in the range of 20 to 80.
2. The leader mixes two sets of Digit Squares and places the cards in an opaque container.
3. The leader draws, announces, and displays one digit at a time. Displayed digits are not returned to the container until the end of the round.
4. Independently, each student decides where to place the drawn digit and records that number in one of the squares.
5. Steps 3 and 4 are repeated until ten digits have been drawn and displayed. Any one digit may be rejected after it is drawn. It is written in the "Reject" square instead of an equation square.

   36.4 + 9.57 - 22.7 = 23.27

   Reject 6   30 Target #   6.73 Difference

6. Answers to the resulting expressions are computed and recorded.
7. Next, students find and record the difference between their own answers and the target number.
8. The Digit Squares are reused for any future rounds. After several rounds, students add their differences to find their scores. The student with the lowest score wins.

Making Connections

Promote reflection and make mathematical connections by asking:

- Which boxes did you try to fill in first? Why?
- Where did you usually place 5s?

Decimal Draw

Recording Sheet

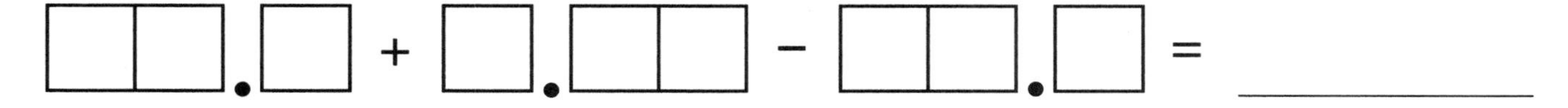

Reject

Target #

Difference

+ − =

Reject

Target #

Difference

+ − =

Reject

Target #

Difference

+ − =

Reject

Target #

Difference

Mystery Factor

Topic: Mental Estimation and Computation

Object: Display target amount using mental division and calculator's constant feature.

Groups: Whole class or small group

Materials

- transparency of *Mystery Factor* recording form, p. 69
- overhead calculator with constant feature
- calculator for each pair (*optional*)

Tips *Students benefit by seeing what products result from estimated factors. Be sure your calculator has a constant feature and take time beforehand to determine if the first or second factor is the constant.*

Directions

1. The leader creates two or three multiplication problems that require a decimal factor to produce a target product. The whole-number factor is set as a constant for the first problem. On many calculators this can be achieved by keying 29 [×] [=].
2. After the leader announces and displays one whole number and a product, the students are asked to estimate the mystery factor.

   Example: 29 × _?_ = 500
3. As each student volunteers a possible mystery factor, the leader uses the constant feature to display the resulting product by keying the announced factor and the equals sign. The resulting product is recorded and displayed for students' reference. Use of the calculator is allowed only when a student "publicly" suggests a possible factor. This estimation process continues until the target whole number is displayed. Decimal representations are expected.
4. Once students experience this warm-up, the class can compete as two teams to determine which group displays the closest mystery factor first.

   Possible starters: 26 × _?_ = 500; 37 × _?_ = 700; × 27 × _?_ = 800; 78 × _?_ = 1000; 33 × _?_ = 1400; 41 × _?_ = 2500; 47 × _?_ = 3000; 53 × _?_ = 7400
5. If desired, two students can share a calculator and take turns displaying the target number. Another options is for student pairs to collaborate to identify the mystery factor. Again, students are allowed to use only the calculator's constant feature.

Known Factor		Mystery Factor		Target Product
29	×	?	=	500
29	×	20	=	580
		16	=	464
		18	=	522
		17	=	493
		17.3	=	501.7
		17.2	=	498.8
		17.25	=	500.25

Making Connections

Promote reflection and make mathematical connections by asking:

- What strategy did you use to identify the unknown decimal factor?

Mystery Factor

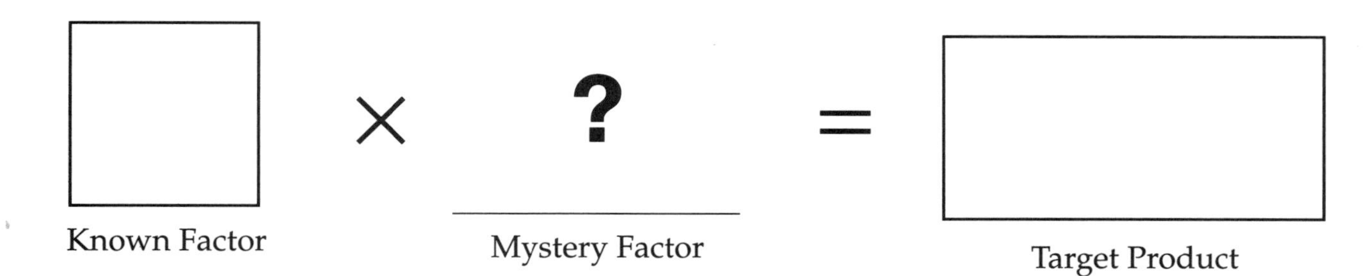

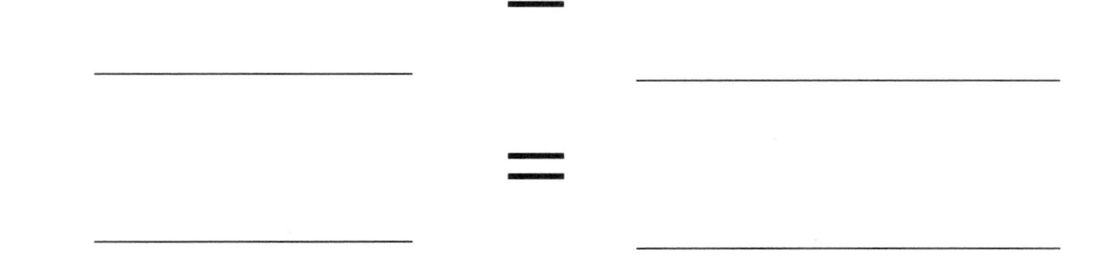

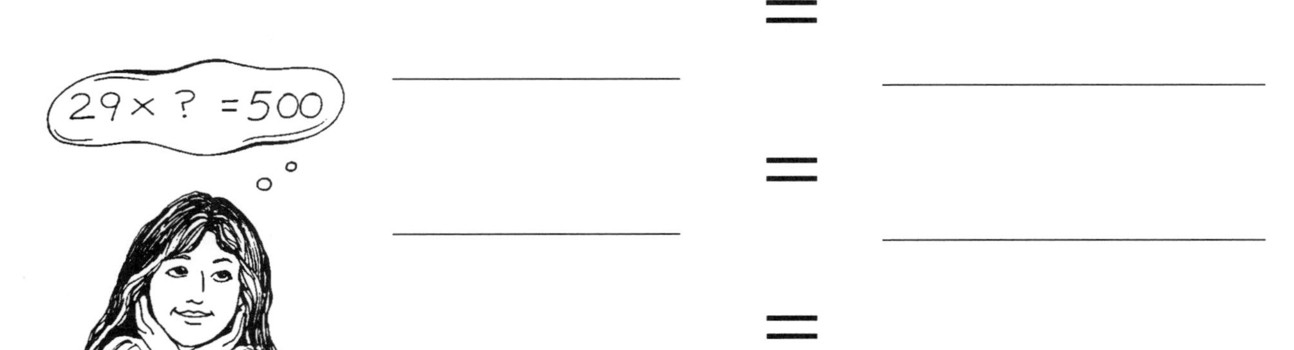

Decimal Tic-Tac-Toe

Topic: Converting Fractions to Decimals

Object: Create a tic-tac-toe.

Groups: Whole class or small group

Materials

- transparency of *Decimal Tic-Tac-Toe* form, p. 71
- *Decimal Tic-Tac-Toe* form for each student
- transparent Digit Squares (1, 2, 3, 4, 6, and 8 only) p. 147
- 9 markers for each student

Directions

1. Students create a unique playing board by randomly recording each listed decimal choice to their *Decimal Tic-Tac-Toe* 3 x 3 playing area. Since only eight choices are listed, each student independently repeats one decimal choice to place in the blank square.
2. The leader draws two Digit Squares from the container. The drawn digits are displayed on the activity form as a fraction, using the lesser number as the numerator.

   Example: 6 and 2 are drawn and displayed as $\frac{2}{6}$.
3. Students individually convert the displayed fraction to a decimal and cover the corresponding decimal with a marker.
4. Before drawing two more Digit Squares, the leader returns the previously drawn Digit Squares to the container.
5. When students have three markers in a row, they call out, "Tic-Tac-Toe." After students share their decimal tic-tac-toes, some students might challenge the results and request proof. The activity continues until every student creates at least one tic-tac-toe.
6. Students can use their created playing boards for additional rounds or create new playing boards.

$\frac{2}{6}$

$0.1\overline{6}$	$0.6\overline{6}$	$0.3\overline{3}$
0.375	0.5	0.25
0.125	0.75	0.5

Tips *If students experience difficulty converting the fractions, allow students to work with a partner in the early rounds. After experiencing a few rounds, students appreciate playing with a 4 x 4 playing area and reusing the same decimal choices.*

Making Connections

Promote reflection and make mathematical connections by asking:

- Which decimal choices were good ones to place twice on your playing board?
- Which digits did you prefer as denominators? Please explain.

Decimal Tic-Tac-Toe

Choices:

0.125	**0.375**
0.1$\overline{6}$	**0.5**
0.25	**0.$\overline{66}$**
0.$\overline{33}$	**0.75**

Date ____________ Name ____________

Partial Possibilities 7

STOP Don't start yet! Star a problem that may have the greatest answer.

1. Write the number four and twenty-three thousandths. ________

2. Write a decimal that comes between: 0.28 ________ 0.29

3. Place the decimal point in each factor.

2 4 5 × 1 6 = 3.92

4. Order from least to greatest.

0. 37, $\frac{1}{4}$, 0.62, $\frac{3}{5}$ ________________

5. Find the approximate value of *B*. ________

A (1.5) ———— B ——— C (2.5)

6. 5.39 × 10 = ________

7. 506.4 + 9.37 – 15.8 = ________

8. Place $<$ or $>$ in the circle.

28.4 × 0.5 ◯ 22 – 6.9

9. 56.4 × 0.72

10. 6.5)280.8

Go On What comes next? 7.69, 12.07, 16.45, ______, ______, ______
Describe the pattern.

Date ____________ Name ____________

Partial Possibilities 8

STOP Don't start yet! Star a problem that may have the least answer.

1. Write the number sixteen and thirty-seven thousandths. ________

2. Write a decimal that comes between: 0.512 ________ 0.513

3. Place the decimal point in each factor.

3 7 1 × 8 2 = 304.22

4. Order from least to greatest.

$\frac{1}{3}$, 0.85, $\frac{4}{5}$, 0.4 ________________

5. Find the approximate value of *B*. ________

6. 7.28 × 100 = ________

7. 213.52 – 28.6 + 5.74 = ________

8. Place $<$ or $>$ in the circle.

24.7 × 0.3 ◯ 64 – 58.9

9. 7.46 × 3.8

10. 4.8)363.84

Go On Use the digits 2, 4, 5, 7, and 8 and two decimal points to write a multiplication problem with the least possible product. Describe your strategy.

Date ______________ Name ______________________

Partial Possibilities 9

Don't start yet! Star a problem that may have an answer in the thousandths.

1. Write the number eighteen and nine thousandths. __________

2. Write a decimal that comes between: 0.43 ________ 0.44

3. Place the decimal point in each factor.

6 3 2 × 4 8 = 30.336

4. Order from least to greatest.

$\frac{2}{4}$, 0.67, $\frac{1}{5}$, 0.17 ________________

5. Find the approximate value of *B*. ________

A (2) — *B* — *C* (3.5)

6. 4.59 × 0.1 = ____________

7. 328.7 + 6.84 – 3.06 = ____________

8. Place < or > in the circle.

36.2 × 0.5 ◯ 57 – 39.6

9. 76.3 × 0.55

10. 0.27)15.228

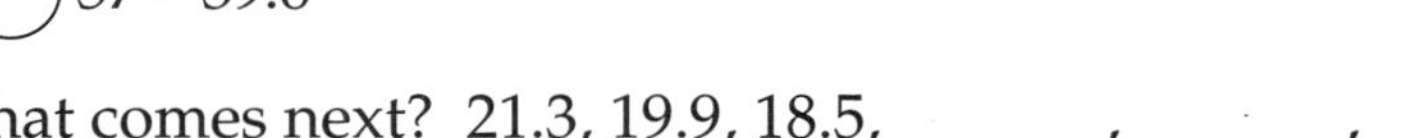

What comes next? 21.3, 19.9, 18.5, ______, ______, ______

Describe the pattern.

Date ______________ Name ______________________

Partial Possibilities 10

Don't start yet! Star a problem that may have the least answer.

1. Write the number three and eighty-one thousandths. __________

2. Write a decimal that comes between: 0.838 ________ 0.837

3. Place the decimal point in each factor.

4 8 × 9 7 5 = 4.68

4. Order from least to greatest.

0.51, $\frac{2}{3}$, 0.65, $\frac{3}{4}$ ________________

5. Find the approximate value of *B*. ________

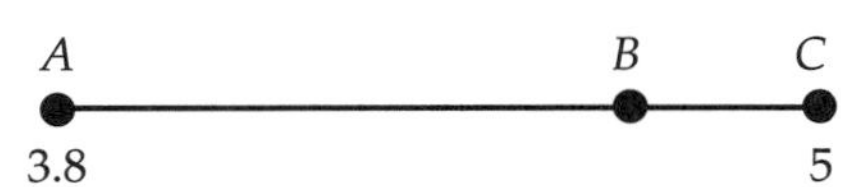

A (3.8) — *B* — *C* (5)

6. 6.12 × 10 = ____________

7. 36.17 + 425.6 – 142.8 = ____________

8. Place < or > in the circle.

63.8 × 0.5 ◯ 53 – 39.1

9. 4.73 × 5.8

10. 7.6)377.872

Go On: Use the digits 1, 3, 6, and 7 and one decimal point to write a division problem with the greatest possible quotient. Describe your strategy.

Date ____________ Name ____________________

Partial Possibilities 11

Don't start yet! Star two problems that may have answers between 3 and 10.

1. Write the number nine and fifty-two thousandths. ________

2. Write a decimal that comes between 0.64 ________ 0.65

3. Place the decimal point in each factor.

7 4 × 3 6 5 = 27.01

4. Order from least to greatest.

0.80, $\frac{7}{8}$, 0.73, $\frac{5}{6}$ ________________

5. Find the approximate value of *B*. ________

A (3) — *B* — *C* (5.4)

6. 18.76 × 100 = ________

7. 453.07 – 60.8 + 9.8 = ________

8. Place $<$ or $>$ in the circle.

52.7 × 0.5 ◯ 42 – 21.6

9. 63.8 × 0.46

10. 0.38$\overline{)2.736}$

Go On What comes next? 2.39, 4.78, 9.56, ______, ______, ______
Describe your strategy.

Date ____________ Name ____________________

Partial Possibilities 12

Don't start yet! Star a problem that may have the greatest answer.

1. Write the number seventeen and sixteen thousandths. ________

2. Write a decimal that comes between: 0.266 ________ 0.267

3. Place the decimal point in each factor.

9 6 1 × 5 3 = 509.33

4. Order from least to greatest.

$\frac{2}{5}$, 0.52, $\frac{1}{3}$, 0.37 ________________

5. Find the approximate value of *B*. ________

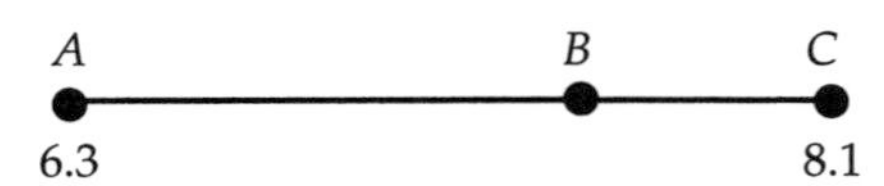

6. 34.8 × 0.1 = ________

7. 162.4 – 8.76 + 9.6 = ________

8. Place $<$ or $>$ in the circle.

36.2 × 0.3 ◯ 47 – 32.9

9. 5.87 × 0.62

10. 5.8$\overline{)1824.1}$

A — *B* — • — • — *C* — *D* — *E*

Go On If *A* = 0 and *E* = 3, what is the value of *B*? of *C*? of *D*?
If *A* = 1 and *E* = 2.8, what is the value of *B*? of *C*? of *D*?

Decimal Four-in-a-Row

Topic: Mentally Multiplying Decimals

Object: Cover four numbers in a row.

Groups: 2 players

Tip *Introduce this engaging game with a transparency of the gameboard and half the class playing against the other half.*

Materials for each group

- *Decimal Four-in-a-Row* gameboard, p. 76
- 2 paper clips
- different kind of markers for each player

Directions

1. The first player places two paper clips at the bottom of the gameboard, indicating two factors. The same player multiplies the selected factors and places a marker on the resulting product.
2. The other player moves only one of the paper clips to a new factor, multiplies the two factors, and places a marker on that product. It is permissible to have two paper clips on the same factor.
3. Players continue alternating turns, moving one paper clip each time, multiplying the factors and placing a marker on the product.
4. The winner is the first player to have four markers in a row horizontally, vertically, or diagonally.

0.01	0.02	0.03	0.04	0.05	0.06
0.07	0.08	0.09	0.1	0.12	0.14
0.15	0.16	0.18	0.2	0.21	0.24
0.25	0.27	0.28	0.3	0.32	0.35
0.36	0.4	0.42	0.45	0.48	0.49
0.54	0.56	0.63	0.64	0.72	0.81

0.1 0.2 0.3 0.4 0.5 0.6 0.7 0.8 0.9

Making Connections

Promote reflection and make mathematical connections by asking:

- What strategies helped you line up your markers in a row?

Decimal Four-in-a-Row

0.01	0.02	0.03	0.04	0.05	0.06
0.07	0.08	0.09	0.1	0.12	0.14
0.15	0.16	0.18	0.2	0.21	0.24
0.25	0.27	0.28	0.3	0.32	0.35
0.36	0.4	0.42	0.45	0.48	0.49
0.54	0.56	0.63	0.64	0.72	0.81

0.1 0.2 0.3 0.4 0.5 0.6 0.7 0.8 0.9

Decimal Bingo

Topic: Mental Division, Decimal Quotients

Object: Cover four numbers in a row.

Groups: 2 players

Materials for each group

- *Decimal Bingo A* or *B* gameboard, pp. 78–79
- different kind of markers for each player
- calculator

Tip *If time is short or when students first learn the game, allow three in a row for a win.*

Directions

1. The first player selects and announces a dividend and a divisor from each choice box. Once the choices are announced, players may use a calculator or pencil and paper to find the quotient. The first player covers the resulting quotient on the gameboard with his or her marker. When necessary, quotients are rounded to the closest hundredth.
2. The second player selects and announces a dividend and a divisor. The quotient can then be found with a calculator or pencil and paper. If the resulting quotient is not already covered, it is covered by that player's marker.
3. Players continue announcing dividends and divisors, and finding quotients, as they alternate turns.
4. The first player to have four markers in a row horizontally, vertically, or diagonally wins.

Dividend Choices

50	110	200	325	450

Divisor Choices

1.5	4.5	5	9	12

24.44	50	9.17	216.67	33.33
22	22.22	40	300	72.22
11.11	133.33	16.67	36.11	100
10	4.17	73.33	90	12.22
5.56	44.44	65	37.5	27.08

Making Connections

Promote reflection and make mathematical connections by asking:

- What strategy helped you correctly estimate quotients?
- What strategy helped you line up your markers in a row?

Decimal Bingo A

Dividend Choices

50	110	200	325	450

Divisor Choices

1.5	4.5	5	9	12

24.44	50	9.17	216.67	33.33
22	22.22	40	300	72.22
11.11	133.33	16.67	36.11	100
10	4.17	73.33	90	12.22
5.56	44.44	65	37.5	27.08

BINGO! Four in a row!

Decimal Bingo B

Dividend Choices

65	140	350	468	696

Divisor Choices

3	5.1	8	12	40

46.67	156	11.67	17.4	3.5
1.63	8.13	12.75	39	116.67
68.63	87	27.45	17.5	8.75
58.5	21.67	5.42	136.47	58
91.76	11.7	232	29.17	43.75

Date ______________________ Name ______________________________________

Decimal Trails I

Find a trail that produces correct answers and alternates the operations of addition and subtraction. Every cell should be included and used only once.

Example:

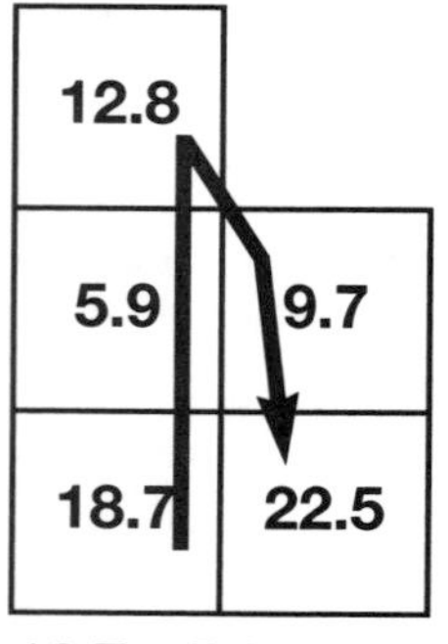

18.7 – 5.9 = 12.8
12.8 + 9.7 = 22.5

1.

324.7	
8.11	313.31
332.81	19.5

2.

39.9	
29.2	69.1
20.5	48.6

3.

30.2	
32.2	62.4
26.9	59.1

4.

75.47	60.53	31.83
28.15	14.94	35.88
47.32	52.77	67.71

5.

24.8	71.1	21.9
46.3	85.7	63.8
49.9	96.2	32.4

6.

35.67	57.48	93.15
43.56	27.88	65.27
79.23	21.75	57.48

7.

32.64	28.5	61.14
58.79	36.64	24.5
95.43	42.5	52.93

8.

211.7	31.42	88.8
234.7	74.48	120.22
266.12	128.9	137.22

9.

75.12	15.4	59.72
48.7	26.42	23.8
42.82	40.7	83.52

10.

280.5	354.37	126.8
73.87	227.57	311.17
38.6	112.47	83.6

11.

67.13	24.4	42.73
19.64	29.8	72.53
47.49	33.81	81.3

12.

53.87	100.7	62.7
154.57	93.97	163.4
60.6	127.6	35.8

Date ____________ Name ____________

Decimal Trails II

Find a trail that produces correct answers and alternates the operations of multiplication and division. Every cell should be included and used only once.

Example:

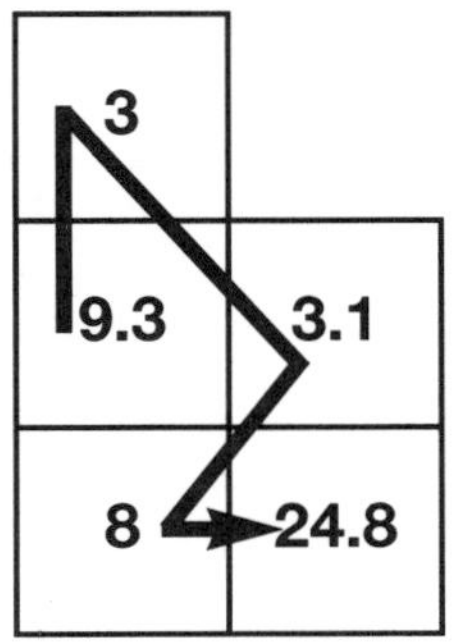

$9.3 \div 3 = 3.1$
$3.1 \times 8 = 24.8$

1.

5	
33	6.6
4.4	7.5

2.

2.5	
13.16	32.9
7	4.7

3.

15.2	
4	3.8
24.7	6.5

4.

2.4	3	7.2
1.5	1.8	4
2.7	3	.9

5.

56.78	3.4	16.7
17	8.35	3.4
3.34	8.5	28.39

6.

5	3.6	3.2
2.7	18	5.625
5	13.5	2.4

7.

103.5	0.75	80
60	6.9	1.8
15	9.6	144

8.

15.8	42.66	12
2.7	4.32	.36
3.2	8.64	24

9.

30.6	4.5	6.8
85	12.5	63.75
20	4.25	15

10.

47	1.5	70.5
4.2	197.4	9.4
27	3.6	7.5

11.

4.4	6.8	135.3
12.3	0.4	29.92
54.12	8.8	3.4

12.

87.78	11	7.98
3.3	8.5	67.83
26.6	1.8	47.88

Date ______________ Name ______________________

Fitting Decimals I

In each sentence, each digit may be used only once. Solve each problem.

1. Use **5, 2, 7, 4,** and **6** to create the greatest possible answer.

□□.3□ + □.□ = ________ □8.□□ – □.□ = ________

□.□□ + □□.5 = ________ □□.□ – 9.□□ = ________

2. Use **8, 6, 4, 9,** and **3** to create the smallest possible answer.

□.□5 + □□.□ = ________ □□.□ – □.7□ = ________

0.□□□ + □2.□ = ________ □□.1 – □8.□□ = ________

3. Use **3, 7, 5, 2,** and **8** to create the smallest possible answer.

□□.9□ + □.□ = ________ □□.□ – 6.□□ = ________

□.1□ + □□4.□ = ________ □.9□□ – □.□ = ________

4. Use **7, 1, 6, 9,** and **3** to create the greatest possible answer.

0.□5□ + □.□□ = ________ □.□ – 0.□□□ = ________

□.4□ + □□.8□ = ________ □2.□ – □□.0□ = ________

5. Use **6, 9, 4, 2,** and **8** to create the smallest possible answer.

□.□□ + 0.7□□ = ________ □□.□ – □.□ = ________

□□.□ + □5.3□ = ________ □1.□ – □□.8□ = ________

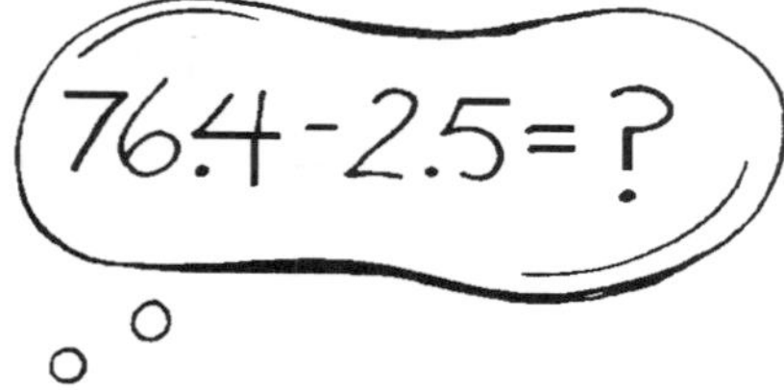

Date ____________ Name ____________________

Fitting Decimals II

In each sentence, each digit may be used only once. Round each answer to the nearest hundredth.

1. Use **4, 1, 8, 5,** and **6** to create the greatest possible answer.

□□.□ × 0.□□ = ________ □.□□ ÷ □.□ = ________

□3.□□ × □.□ = ________ □□.□ ÷ 6.□□ = ________

2. Use **3, 7, 4, 9,** and **5** to create the smallest possible answer.

□6.□□ × □.□ = ________ □□.7□ ÷ 0.□□ = ________

8.□□□ × □.2□ = ________ □6.□□ ÷ □.□ = ________

3. Use **8, 2, 5, 6,** and **3** to create the smallest possible answer.

4.□.□ × □.□□ = ________ □□□ ÷ 0.□□ = ________

9.□ × □□.□□ = ________ □□.7□ ÷ □.□ = ________

4. Use **5, 7, 2, 9,** and **4** to create the greatest possible answer.

6□.□ × 0.□□□ = ________ 3□.□□ ÷ 0.□□ = ________

□1.9□□ × □.□ = ________ □.□□ ÷ 0.□□ = ________

5. Use **7, 3, 1, 6,** and **4** to create the smallest possible answer.

□.□□ × 0.5□□ = ________ □8□.□ ÷ □.□ = ________

9□.4□ × □.□□ = ________ □1.□□ ÷ 0.□□ ________

Date ____________________ Name ________________________________

Magic Squares

In a magic square, the sum of the numbers in each row, column and diagonal is the same. Find the missing numbers to create magic squares. Good luck!

180.2		194
	164.4	
		148.6

	168.1	
263.7		
193.8	315.1	215.9

		357.7
		294.6
473.3		594.2

32.9		39.8	115.7
108.8		74.3	26
	53.6		
12.2			95

			211.2
	159.3		
142	90.1	107.4	193.9
55.5		228.5	3.6

17.4		195.2	55.5
	93.6	80.9	
		131.7	68.2
169.8		42.8	

6.1				36.1
	19.8	58.3	34.8	4.8
		33.5	3.5	
62.2			47.2	17.2
30.9	7.4	45.9		60.9

Percents

Assumptions Percent concepts and calculations have previously been taught and reviewed, emphasizing understanding and building number sense. Concrete objects and visual models, such as base-ten materials and grids, have been used extensively.

Section Overview and Suggestions

Sponges

What's My Word? p. 86

Fractions to Percents pp. 87–88

Ordering Percents and Fractions p. 89

Exploring Percent Relationships pp. 90–91

These repeatable whole-class or small-group warm-ups build percent and number sense as students enhance their abilities to convert and compute with percents mentally and with paper and pencil. Frequent and repeated use of *Fractions to Percents* will ensure greater success with all the Games and Independent Activities in this section.

Skill Checks

Perky Portions 1–6 pp. 92–94

These provide a way for parents, students, and you to see students' improvement with percent conversion and computation. Copies can be cut in half so that each Check may be used at a different time. Remember to have all students respond to STOP, the number-sense task, before solving the ten problems.

Games

Respond and Travel pp. 95–97

Line Up Four pp. 98–99

Ordered Pathways pp. 100–101

These open-ended and repeatable Games actively involve students in converting and comparing percents, fractions, and decimals as they enhance their strategic thinking abilities. To ensure more active and successful participation with *Ordered Pathways,* provide adequate practice with the *Ordering Percents and Fractions* Sponge.

Independent Activities

Ordered Mazes p. 102

Relating Quantities p. 103

Cross-Number Puzzle p. 104

Each Independent Activity requires students to practice converting and comparing percents, fractions, and decimals. These challenging activities involve mental computation and lead to improved number sense.

What's My Word?

Topic: Finding Percents, Fractions, and Decimals

Object: Use percent, fraction, and decimal clues to identify a secret word.

Groups: Whole class or small group

Materials

- prepared clues (use examples in #1 and #2 below as starters)

Directions

1. The leader orally states two words which volunteers display for classmates to see. The leader provides percent, fraction, or decimal clues and a description of the secret word. Students use parts of the two words to form a new word that matches the clue.

   Example: "Use 25% of *fraction* and $\frac{2}{3}$ of *quench* to identify a language." (Fr + ench = French)

2. Additional words and clues are created following the Example above. Some students will benefit from having written and displayed clues.

   Additional examples:

   Use 50% of *spring* and $\frac{1}{3}$ of *allowance* to identify the son of a king. (prince)

   Use $\frac{1}{3}$ of *dachshund* and 40% of *spinnakers* to identify a vegetable. (spinach)

   Use $\frac{2}{9}$ of *recognize* and 0.4 of *staff* to identify a popular fast food. (taco)

   Use $\frac{17}{34}$ of *astonish* and 30% of *thoughtful* to identify a city. (Houston)

3. After students solve several puzzles, have them work in pairs to create puzzles for classmates to solve. Students should have access to dictionaries.

Making Connections

Promote reflection and make mathematical connections by asking:

- Was it easier to create or solve these puzzles? Which procedure was more enjoyable?

FRaction + quENCH =
What language?

Tips *If students experience difficulty, begin by providing more specific information, such as the last 25% of* dynamite. *The sequenced examples provide a model for making this warm-up increasingly more difficult.*

Fractions to Percents

Topic: Converting Fractions to Percents

Object: Create equivalent fractions and percents.

Groups: Pair players within whole class or small groups

Tip *You may wish to begin at an easier level by excluding the 7 and 9 Digit Squares.*

Materials

- *Fractions to Percents* recording sheet for each pair of students, p. 88
- transparent set of Digit Squares (0 removed), p. 147
- set of Digit Squares for each pair of students

Directions

1. The leader draws and announces the identity of two Digit Squares. The Digit Squares are displayed as a common fraction—the greater digit in the denominator.

Example: For 8 and 3, the fraction $\frac{3}{8}$ is displayed.

2. Pairs place the two drawn digits in the fraction frame at the top of their *Fractions to Percents* recording sheets. Then they collaborate to generate and record three equivalent fractions.

=	$\frac{6}{16}$	=	$\frac{12}{32}$	=	$\frac{15}{40}$	=	$\frac{36}{96}$	=	37.5 %
	Fraction		Fraction		Fraction		Hundredths or Close		

3. Next, each pair records a fourth equivalent fraction with a denominator close to or exactly 100. Some students may prefer to refer to one of the other fraction equivalents.

Example: If $\frac{3}{8}$ is displayed, $\frac{36}{96}$ is a fraction with a denominator close to 100.

4. Each pair determines and records the percent that is equivalent to or nearly equivalent to the original fraction.

5. The leader has students volunteer a variety of fraction equivalents for the original fraction. When no more fraction equivalents can be found, ask a student to state the percent equivalent and justify it.

6. After the transparent Digit Squares are returned and mixed, two new Digit Squares are drawn and the process is repeated with a new fraction. The recording sheet allows five rounds of this Sponge.

Making Connections

Promote reflection and make mathematical connections by asking:

- What helped you to mentally determine the percent equivalents?

Fractions to Percents

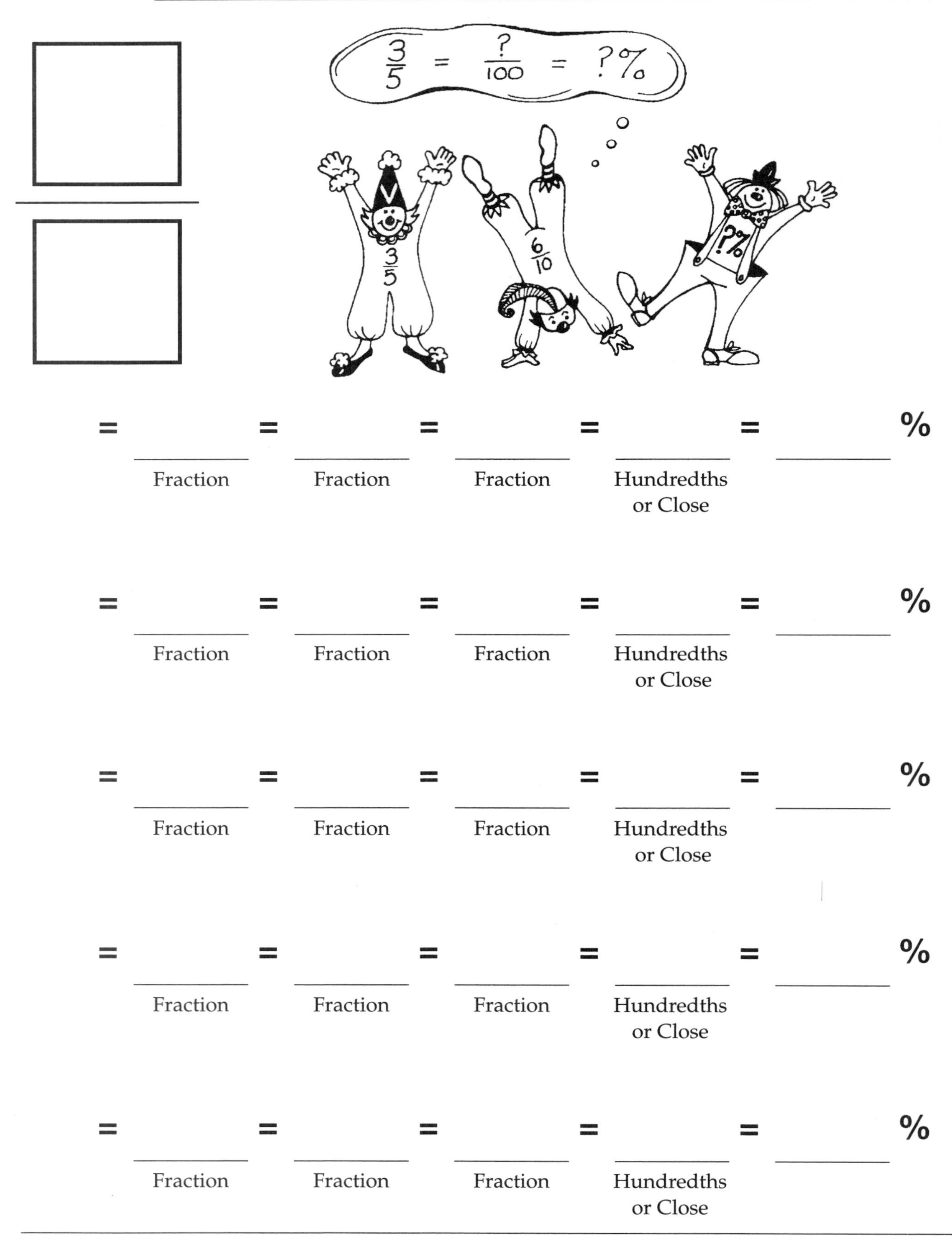

= ____ = ____ = ____ = ____ = ____ %

Fraction | Fraction | Fraction | Hundredths or Close

= ____ = ____ = ____ = ____ = ____ %

Fraction | Fraction | Fraction | Hundredths or Close

= ____ = ____ = ____ = ____ = ____ %

Fraction | Fraction | Fraction | Hundredths or Close

= ____ = ____ = ____ = ____ = ____ %

Fraction | Fraction | Fraction | Hundredths or Close

= ____ = ____ = ____ = ____ = ____ %

Fraction | Fraction | Fraction | Hundredths or Close

Ordering Percents and Fractions

Topic: Comparing Values of Percents and Fractions

Object: Order a mixture of fractions and percents from least to greatest.

Groups: Whole class or small group

Tips *If time allows, have teams play another round with the other team having the second turn. Increase difficulty by requiring use of all the displayed fractions and percents.*

Materials

- transparent set of Digit Squares (0 removed), p. 147

Directions

1. The leader draws and displays three Digit Squares. Students use these digits to express different whole percents. As each possibility is mentioned, it is recorded and displayed.

   Example: For 2, 3, and 8 some possibilities are 2%, 23%, 38%, 82%, and 382%.

2. Next students use the displayed digits to express a variety of fractions and mixed numbers. These are listed and displayed.

   Some possible fractions: $\frac{2}{3}$, $\frac{2}{8}$, $\frac{3}{8}$, $\frac{3}{2}$, $\frac{8}{3}$, $2\frac{3}{8}$, and $\frac{8}{2}$.

3. The group is divided into two teams. The leader displays ten horizontal lines in a row and announces that eventually ten of the displayed numbers need to be recorded in order from least to greatest.
4. The first team selects one of the displayed percents or fractions, crosses out the selected amount, and records the amount on one of the ten lines.
5. The other team chooses one of the remaining percents or fractions and records it on another line. If a team does not follow the least to greatest order, that team loses its turn and the other team receives a point.

   ___ $\frac{2}{8}$ $\frac{3}{8}$ ___ ___ ___ $2\frac{3}{8}$ ___ 382% ___

   least ... greatest

6. If a team is able to prove that none of the remaining lines can be filled by the remaining numbers, that team receives one point and the round ends. This scoring system is designed to encourage collaboration.
7. If the teams are able to correctly order ten different displayed amounts, each team receives a point and the team that records the tenth number receives a bonus point.

Making Connections

Promote reflection and make mathematical connections by asking:

- What helped you order the fractions with the percents?
- What strategy works well to ensure that all ten lines will be used?

Exploring Percent Relationships

Topic: Finding Percents and Related Amounts

Object: Calculate with percents, and justify solutions.

Groups: Whole class or small group

Tips *Have students help generate additional problems for the group to solve. When students gain confidence with this activity, restrict use of paper and pencil and require them to solve posed percent problems mentally.*

Materials

- prepared percent problems, p. 91

Directions

1. The group is divided into two teams.
2. The leader displays a percent problem, which is independently solved by each student. The leader asks students to also prepare an explanation to justify their solution.

   Possible problems: 25% of 76; $33\frac{1}{3}$% of 36; 50% of 154; 90% of 80
3. Students from each team work in pairs to share solutions and explanations. Disagreements are discussed and resolved.
4. A volunteer from each team prepares to present the team's response by recording the solution.
5. The leader asks each team's representative to share the team's solution and provide a convincing justification.
6. If solutions and explanations are acceptable, each team receives one point. Each member of each team is allowed to present a solution and an explanation only once.
7. The process of displaying problems, individually solving problems, sharing with a partner, and presenting solutions and justifications to the entire group is repeated. A leveled sequence of possible problems is provided on the next page. Require students to explain reasoning for estimates. As students gain more competence and confidence, mix the problem types more frequently.
8. After a student from each team has presented a solution and convincing justification to the entire group, the points are totaled and announced.

Making Connections

Promote reflection and make mathematical connections by asking:

- Which explanations increased your understanding of percents?
- Which problems were you able to solve mentally? Explain.

Exploring Percent Relationships

Possible Problems

11% of 40
24% of 198

20% off $60
25% off $150

$6.99 + 6% tax
$12.90 + 7% tax

Estimate: 12% of 63
Estimate: 31% of 150

_____% of 160 = 40
_____% of 500 = 200

50% of _____ = 40
75% of _____ = 180

55% of 150
94% of 120

15% off $24
20% off $14.95

$25 + 7.5% tax
$48 + 7.75% tax

Estimate: 20% of 73
Estimate: 58% of 82

_____% of 175 = 35
_____% of 320 = 40

40% of _____ = 26
55% of _____ = 99

Date ____________ Name ____________

Perky Portions 1

STOP Don't start yet! Star a problem that may have an answer greater than 100.

1. $\frac{3}{5}$ = _____ % **2.** 65% = $\frac{\square}{\square}$ (lowest terms) **3.** 50% of 200 = _____ **4.** 60% of 120 = _____

5. 0.025 = _____% **6.** _____ % of 96 = 16 **7.** 30% of _____ = 54

8. Order from least to greatest: $\frac{1}{6}$, 82%, 0.63, 1.7%

____

9. \$14.50 + 7.5% tax = _____

10. A \$50 savings account earns 6% interest annually.
What is the value of the account after one year? _______________

Estimate 12% of 63 and justify your estimate.

Date ____________ Name ____________

Perky Portions 2

STOP Don't start yet! Star a problem that may have the least answer.

1. $\frac{5}{8}$ = _____ % **2.** 0.38 = $\frac{\square}{\square}$ (lowest terms) **3.** 25% of 320 = _____ **4.** $16\frac{2}{3}$ % of 42 = _____

5. 0.125 = _____% **6.** _____ % of 54 = 27 **7.** 25% of _____ = 10

8. Order from least to greatest: 37%, 0.73, $\frac{3}{7}$, $\frac{7}{3}$

____

9. \$37.25 + 6.5% tax = _____

10. A \$125 savings account earns 7% interest annually.
What is the value of the account after one year? _______________

How do you know that 18% of 36 equals 36% of 18?

Date ______________ Name ______________________

Perky Portions 3

STOP Don't start yet! Star a problem that may have the greatest answer.

1. $\frac{4}{5}$ = _____ % **2.** 85% = $\frac{\square}{\square}$ (lowest terms) **3.** 60% of 40 = _____ **4.** 55% of 160 = _____

5. 0.092 = _____% **6.** _____ % of 200 = 112 **7.** 5% of _____ = 10

8. Order from least to greatest: 0.37, $\frac{3}{8}$, 52%, 0.41 **9.** \$12.98 + 6.5% tax = _____

____

10. A \$80 savings account earns 6% interest annually.
What is the value of the account after one year? ______________________

Go On How do you know that 15% of 512 is greater than 30% of 250?

Date ______________ Name ______________________

Perky Portions 4

Don't start yet! Star a problem that may have an answer between 50 and 100.

1. $\frac{3}{8}$ = _____ % **2.** 0.35 = $\frac{\square}{\square}$ (lowest terms) **3.** 40% of 200 = _____ **4.** 30% of 20 = _____

5. 0.365 = _____% **6.** _____ % of 45 = 36 **7.** 72% of _____ = 108

8. Order from least to greatest: 0.56, $\frac{5}{6}$, 65%, $\frac{5}{8}$ **9.** \$29.98 + 7.5% tax = _____

____

10. A \$175 savings account earns 7% interest annually.
What is the value of the account after one year? ______________________

Go On Estimate 19% of 75 and justify your estimate.

Date ____________ Name ____________________

Perky Portions 5

Don't start yet! Star a problem that may have the greatest answer.

1. $\frac{2}{5}$ = _____ %

2. 45% = $\frac{\square}{\square}$ lowest terms

3. 25% of 180 = _____

4. 80% of 40 = _____

5. 0.087 = _____%

6. _____ % of 150 = 120

7. 40% of _____ = 24

8. Order from least to greatest: $\frac{9}{17}$, 40%, 1.27, $\frac{2}{3}$

____

9. $18.25 + 6.5% tax = _____

10. A $200 savings account earns 6% interest annually.
What is the value of the account after one year? _________

How do you know that 28% of 62 equals 62% of 28?

Date ____________ Name ____________________

Perky Portions 6

Don't start yet! Star a problem that may have an answer close to $\frac{1}{2}$.

1. $\frac{7}{8}$ = _____ %

2. 0.47 = $\frac{\square}{\square}$ lowest terms

3. 80% of 55 = _____

4. $66\frac{2}{3}$% of 150 = _____

5. 0. 762 = _____%

6. _____ % of 75 = 60

7. 30% of _____ = 21

8. Order from least to greatest: 64%, $\frac{8}{17}$, 1.03, $\frac{5}{8}$

____

9. $49.50 + 7.5% tax = _____

10. A $245 savings account earns 7% interest annually.
What is the value of the account after one year? ____________________

How do you know that 30% of 711 is greater than 60% of 325?

Respond and Travel

Topic: Determining Equivalent Percents and Fractions

Object: Create a pathway across the gameboard.

Groups: 2 players or pair players

Tip *Invite interested students to create their own gameboards and accompanying clue cards.*

Materials for each group

- *Respond and Travel* gameboard, p. 96
- *Respond and Travel* Clue Cards, cut apart, p. 97
- different kind of markers for each player

Directions

1. The first player draws a clue card, reads it aloud, and finds an equivalent fraction or percent. If the other player agrees with the response, the first player locates and covers a corresponding cell. If the first player's response is inaccurate, he or she loses that turn.
2. The other player follows the same procedure.
3. Only one marker can occupy a cell.
4. If a player draws a card, and both players agree that there is no uncovered cell with an equivalent fraction or percent, the player discards the card and draws another.
5. Players continue alternating turns until one player wins by forming a continuous pathway across the gameboard, from top to bottom or left to right.

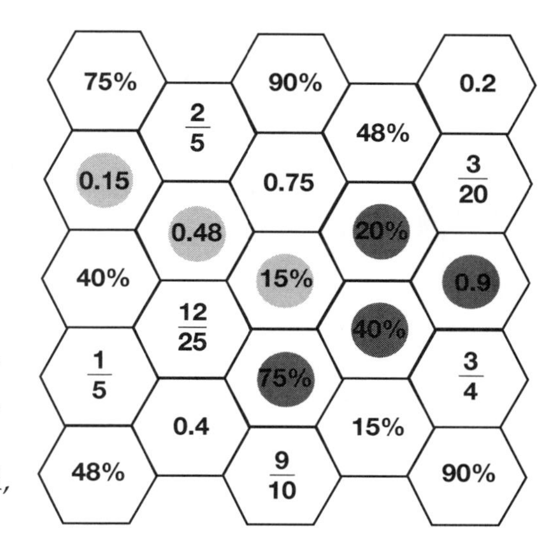

Making Connections

Promote reflection and make mathematical connections by asking:

- Which clue cards were more difficult to answer?
- What strategy helped you place your markers in a complete pathway?

Respond and Travel

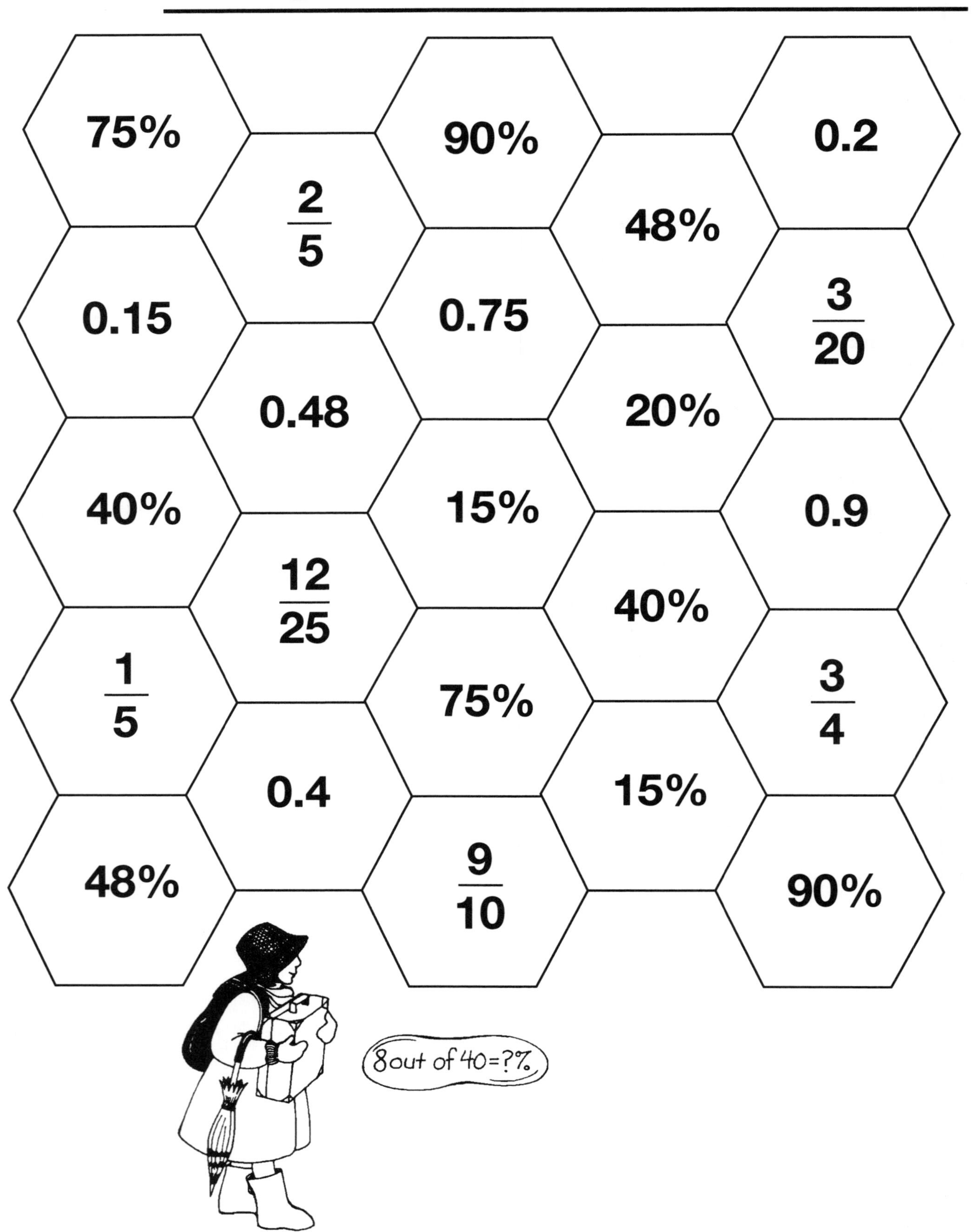

Respond and Travel

Clue Cards

8 out of 40	6 out of 40	24 out of 50	15 out of 20	20 out of 50
36 out of 40	10 out of 25	12 out of 80	11 out of 55	63 out of 70
36 out of 75	37.5 out of 50	32 out of 80	19.8 out of 22	9 out of 60
15 out of 75	66 out of 88	43.2 out of 90	18 out of 90	7.2 out of 15
51 out of 68	28 out of 70	11.25 out of 75	40.5 out of 45	22.4 out of 56
50.4 out of 56	22.5 out of 30	28.8 out of 60	7.2 out of 36	10.5 out of 70

Line Up Four

Topic: Converting Common Fractions to Decimals and Percents

Object: Cover four numbers in a row.

Groups: 2 players or pair players

Materials for each group

- *Line Up Four* gameboard, p. 99
- set of Digit Squares in opaque container (0, 7, and 9 removed), p. 147
- 10 transparent markers, 5 one color and 5 of another

Tip *For a greater challenge, provide six markers for each player and require winners to place five markers in a row.*

Directions

1. The first player draws two Digit Squares and displays them as a common fraction with the greater digit as the denominator. The player locates any equivalent decimal or percent on the gameboard and places one of his or her five markers on that square.
2. The displayed Digit Squares are returned to the container after each player's draw.
3. The second player draws two Digit Squares and repeats the process.
4. The first player takes a second turn following the same procedure. The player has a choice of putting a second marker on an uncovered equivalent or moving a previously placed marker onto that equivalent. If an equivalent is available, a player must cover it.
5. If all available equivalents are covered, a player may replace the other player's marker with one of his or her markers. Removed markers are returned to the other player. Sometimes a player may not have a move, because any equivalent already holds one of his or her markers.
6. Players take turns following the steps outlined above.
7. After all five of a player's markers are placed on the gameboard, the player must select any previously placed marker and move that marker to an available equivalent.
8. The first player to have four markers in a row horizontally, vertically, or diagonally wins.

$33\frac{1}{3}\%$	0.5	62.5%	75%	0.25
$0.1\overline{66}$	25%	$66\frac{2}{3}\%$	50%	0.6
$83\frac{1}{3}\%$	0.75	$0.\overline{33}$	40%	$66\frac{2}{3}\%$
0.375	$0.\overline{66}$	12.5%	0.2	50%
50%	0.25	80%	$0.\overline{33}$	0.75

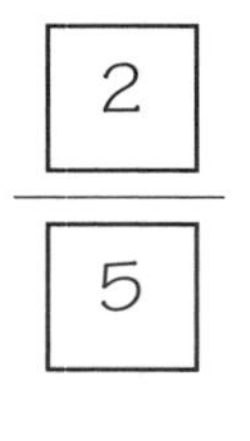

Making Connections

Promote reflection and make mathematical connections by asking:

- Did any player avoid placing his or her five markers on the first five turns? If so, what was the reasoning behind this strategy?
- How would the game be different if the gameboard cells could be shared by the two players' markers?

Line Up Four

$33\frac{1}{3}\%$	0.5	62.5%	75%	0.25
$0.1\overline{66}$	25%	$66\frac{2}{3}\%$	50%	0.6
$83\frac{1}{3}\%$	0.75	$0.\overline{33}$	40%	$66\frac{2}{3}\%$
0.375	$0.\overline{66}$	12.5%	0.2	50%
50%	0.25	80%	$0.\overline{33}$	0.75

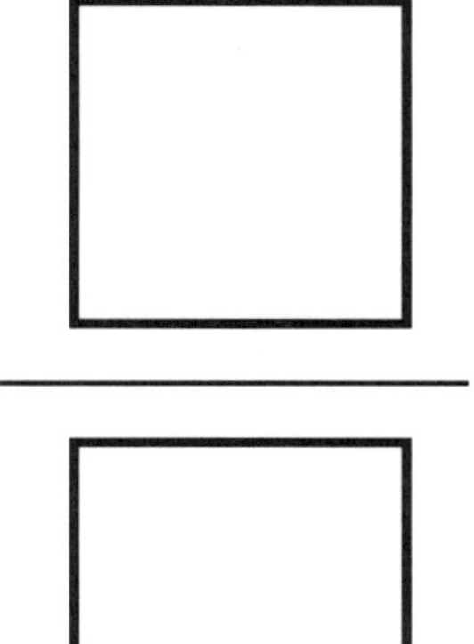

Ordered Pathways

Topic: Comparing Percents, Fractions, and Decimals

Object: Record percents, fractions, and decimals in increasing order.

Groups: 2 players or pair players

Tip *Prepare students for independent success with this game by dividing the class in half and playing this as a team game, with each team displaying a separate recording sheet.*

Materials for each group

- 2 sets of Digit Cards (0 removed), p. 146
- *Ordered Pathways* recording sheet for each player, p. 101

Directions

1. A player mixes two sets of Digit Cards together and stacks them facedown.
2. Each player draws two cards and uses the drawn digits to form a fraction, a decimal, or a percent between 0 and 3.

   Example: If 8 and 2 are drawn, the player might choose 82%, 2.8, $\frac{2}{8}$, or 0.82.
3. Keeping in mind the position of the possible numbers between 0 and 3, each player selects and records his or her choice in a cell along the pathway. After recording their choices, the players share their decisions with each other. If each player accepts his or her opponent's recording, play continues. Each player's pathway must eventually include at least two decimals, two percents, and two fractions. Drawn cards are set aside.
4. Players draw two new Cards and repeat these steps. After four turns, all Digit Cards are mixed and restacked.
5. A player loses a turn if the drawn digits cannot form a fraction, decimal, or percent that will fit in a cell along the ordered pathway.
6. Play continues until one player correctly completes a pathway that includes at least two decimals, two percents, and two fractions.

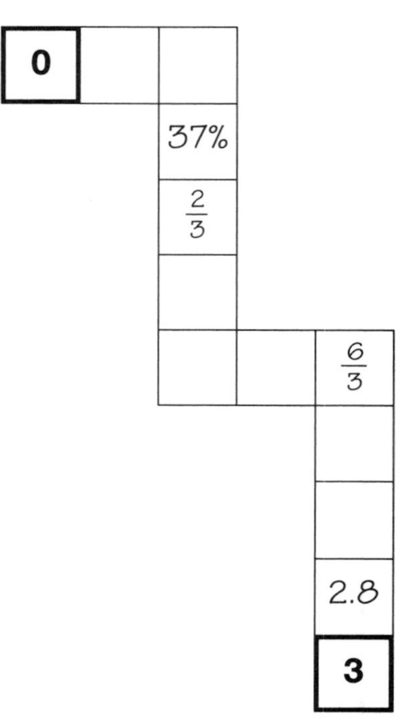

Making Connections

Promote reflection and make mathematical connections by asking:

- Was it easier to place fractions, decimals, or percents? Explain.
- How could you improve the rules of this game?

Ordered Pathways

Date ____________________ Name ______________________________

Ordered Mazes

Shade in a path from START to FINISH. Each number on your path must be greater than the previous one.

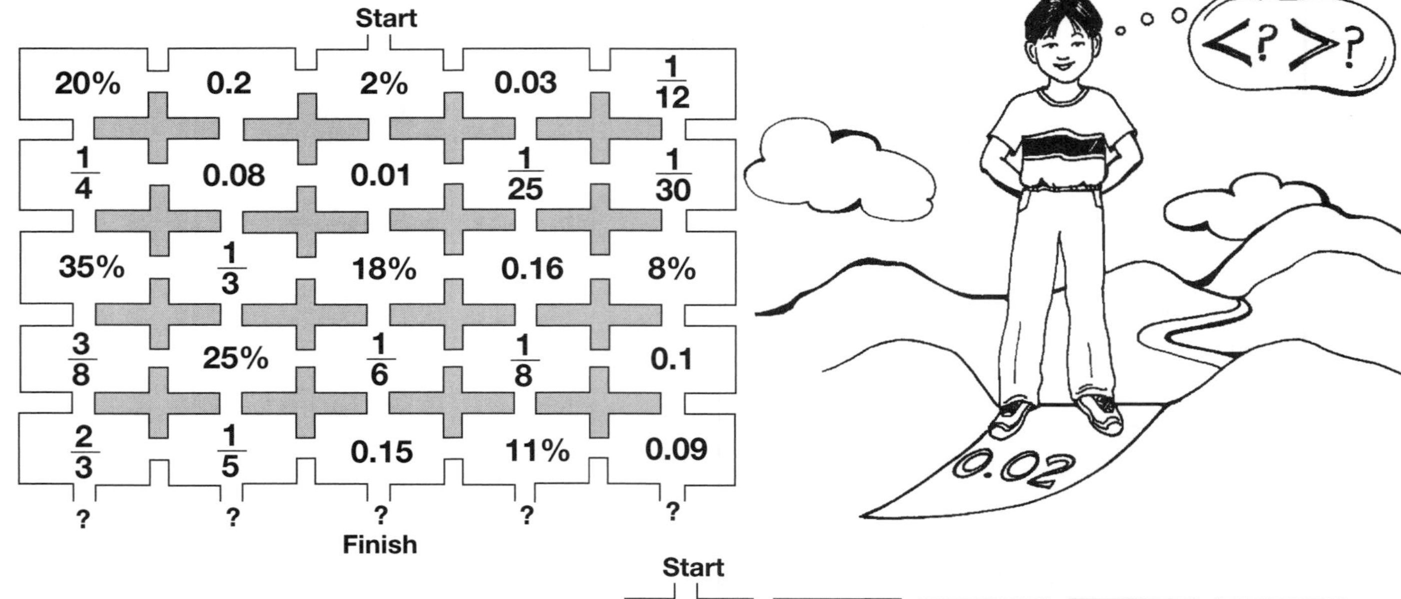

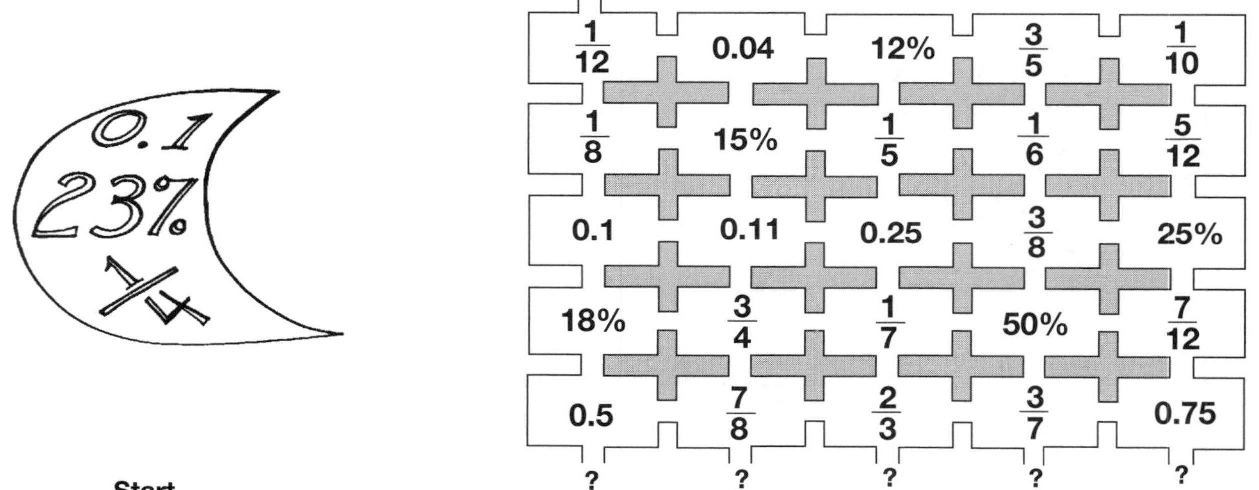

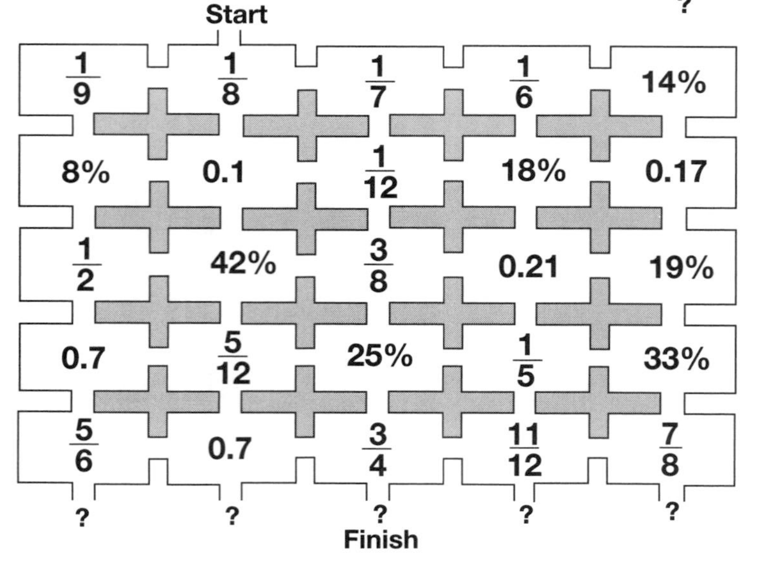

Date ____________ Name ____________________

Relating Quantities

Use one number from each box to complete the problems on the right.

25%	$\frac{3}{5}$	0.1	40%	$\frac{2}{3}$

First order the numbers in the box above from least to greatest.

_____ _____ _____ _____ _____

60	120
80	200
100	300

_____ of _____ = 25　　_____ of _____ = 72

_____ of _____ = 180　　_____ of _____ = 32

_____ of _____ = 80　　_____ of _____ = 200

_____ of _____ = 48　　_____ of _____ = 12

_____ of _____ = 50　　_____ of _____ = 60

Can you do the problems below in two different ways?

_____ of _____ = 30　　_____ of _____ = 30

_____ of _____ = 40　　_____ of _____ = 40

50%	$\frac{4}{5}$	0.35	70%	$\frac{5}{8}$

First order the numbers in the box above from least to greatest.

_____ _____ _____ _____ _____

40	120
60	150
100	200

_____ of _____ = 80　　_____ of _____ = 48

_____ of _____ = 25　　_____ of _____ = 75

_____ of _____ = 140　　_____ of _____ = 28

_____ of _____ = 84　　_____ of _____ = 70

_____ of _____ = 32　　_____ of _____ = 100

Can you do the problems below in two different ways?

_____ of _____ = 75　　_____ of _____ = 75

_____ of _____ = 42　　_____ of _____ = 42

60%	$\frac{2}{5}$	0.2	15%	$\frac{1}{4}$

First order the numbers in the box above from least to greatest.

_____ _____ _____ _____ _____

50	140
75	150
100	200

_____ of _____ = 45　　_____ of _____ = 20

_____ of _____ = 84　　_____ of _____ = 21

_____ of _____ = 10　　_____ of _____ = 40

_____ of _____ = 50　　_____ of _____ = 80

_____ of _____ = 90　　_____ of _____ = 15

Can you do the problems below in two different ways?

_____ of _____ = 30　　_____ of _____ = 30

_____ of _____ = 60　　_____ of _____ = 60

Date ____________________ Name __________________________________

Cross-Number Puzzle

1.	2.			3.		4.		
	5.					6.		7.
8.				9.				
		10.				11.	12.	
13.	14.				15.		16.	
			17.				18.	

Across

1. $20 dinner + 15% tip = $ ______
3. 25% of 500 = _____
5. $30.00 + 7.5% tax = $ __ __.__ __
6. 25% off $728 = $ ______
8. ____% of 25 = 14
9. 24 out of 50 = ____%
10. 80% of 120
11. ____% of 300 = 228
13. 5% off $900 = $ ______
16. _____% of 525 = 294
17. 102 = 75% of _____
18. ____% of 200 = 16

Down

2. 20% off $420 = $ ______
3. _____ out of 200 = 77%
4. ____% of 200 = 110
7. 16 of 25 free shots made = ____%
8. 20% discount on $660 = $ ______
10. 19 out of 20 = ____%
12. 30% off $940 = $ ______
14. _____ out of 295 = 20%
15. Year's interest earned on $320 at 5% annually

Trivia: 10 Down is the percent of a jellyfish that consists of water.

Integers

Assumptions Integer concepts and operations of integers have been taught and reviewed, emphasizing understanding and building integer sense. Two-sided counters and other visual models, such as number lines, have been used extensively.

Section Overview and Suggestions

Sponges

Subtracting Integers pp. 106–107

Negatives Score pp. 108–109

How? pp. 110–111

These repeatable whole-class or small-group warm-ups build integer sense as students demonstrate and enhance their abilities to model integer operations and to mentally compute integers.

Skill Checks

Negative Notions 1–6 pp. 112–114

These provide a way for parents, students, and you to see students' improvement with operations of integers. Copies can be cut in half so checks may be used at different times. Remember to have all students respond to STOP, the number sense task, before solving the ten problems.

Games

Integer Four-in-a-Row pp. 115–116

Cover It! pp. 117–118

Neighboring Integers Count pp. 119–120

These open-ended and repeatable Games actively involve students in computing with integers, while students enhance their strategic thinking abilities. Adequate use of the *Subtracting Integers* Sponge should ensure student success with *Integer Four-in-a-Row,* a Game that emphasizes the more difficult task of subtracting integers. *Cover It!* and *Neighboring Integers Count* reinforce all operations and work well for additional practice at home or family math sessions.

Independent Activities

Finding Sums and Differences p. 121

Finding Tic-Tac-Toes pp. 122–123

Integer Choices p. 124

Each Independent Activity easily engages students in practicing integer operations. These challenging Activities encourage mental computation as students improve their integer sense. *Finding Tic-Tac-Toes* includes two versions, allowing practice of addition and subtraction or multiplication and division. By frequently changing the four choices of integers, the popular *Integer Choices* is a repeatable Activity that provides long-term practice of all operations.

Subtracting Integers

Topic: Subtracting Integers

Object: Demonstrate different ways to solve the displayed equation.

Groups: Whole class or small group

Tip *Initially, have students partner with a classmate to demonstrate justification for solutions.*

Materials

- transparency of *Subtracting Integers* activity form, p. 107
- 2 sets of transparent Negative Integer Squares, p. 148
- 1 set of transparent Digit Squares (0 removed), p. 147
- opaque container
- 2 colors of transparent chips

Directions

1. Negative Integer Squares and Digit Squares are mixed and placed in the container.
2. The leader draws one Square and displays it in the sum box.
3. The leader draws and displays a second Square in the known addend box.
4. Students are asked to individually complete the displayed equation.
5. Students volunteer to justify their solution in a variety of ways. They might use chips of different colors, illustrate on the open-ended number line, provide a missing-numeral explanation, or describe a situation with context.

   Example: Display shows **4** – **⁻3** =

   Using chips of different colors, one student displays 4 positive chips and states that one needs to remove 3 negative chips. To do so, the student adds zero by adding 3 negative chips and 3 positive chips. After removing the 3 negative chips, the student has 7 positive chips.
6. The drawn Squares are returned to the container and the process is repeated.

Making Connections

Promote reflection and make mathematical connections by asking:

- Which models made it easier to justify your solutions? Explain.
- What helps you understand subtracting negative integers?

Subtracting Integers

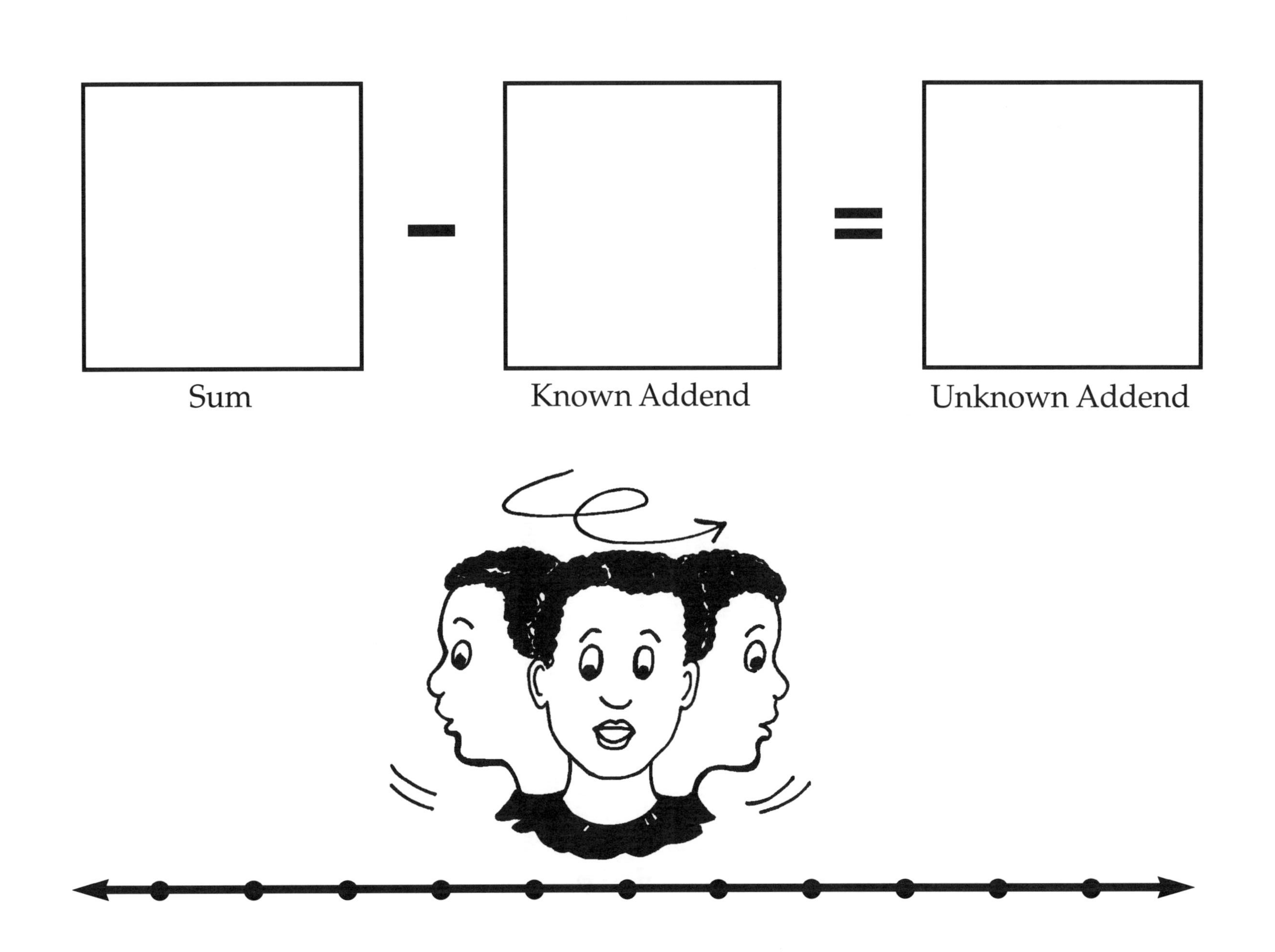

Negatives Score

Topic: Modeling Integer Operations

Object: Compute with both positive and negative integers to yield zero or an integer close to zero.

Groups: Whole class or small group

Tips *Encourage multiplication and division by rewarding an additional point for using these operations. Extend activity to digits 1–9 by drawing from two sets of Digit Squares.*

Materials

- transparency of *Negatives Score* activity form, p. 109
- 6 Number Cubes (1-6)

Directions

1. The class is divided into two teams.
2. A volunteer from each team rolls three Number Cubes. The resulting numbers for each team are displayed on the activity form. The displayed numbers can be used as negative or positive integers.

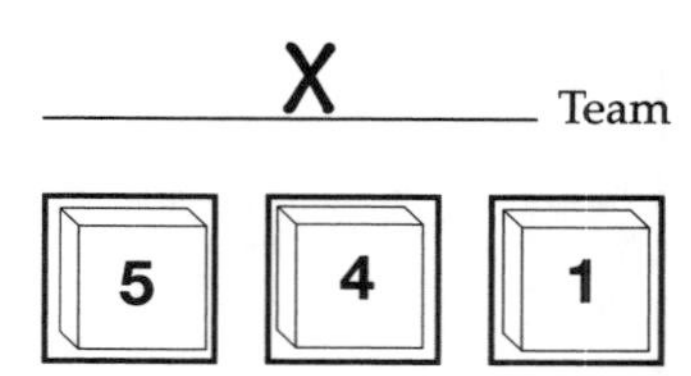

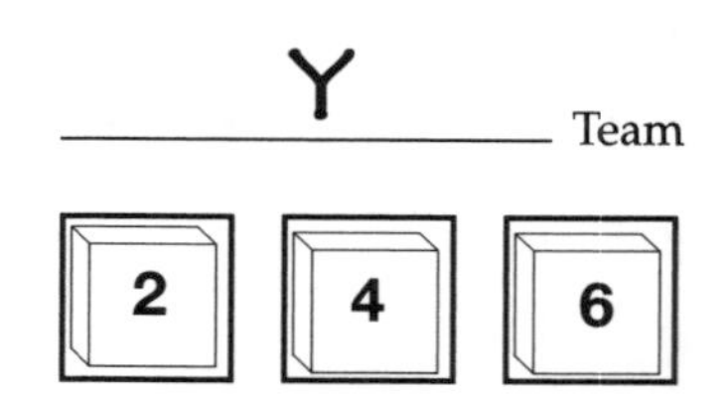

3. Team members collaborate to create computations that include negative numbers and yield zero or an integer close to zero. Students are allowed to use any operation and parentheses.
4. Team members are required to display their solutions on the transparency and justify their solutions on the number line.
5. Qualifying equations receive points. Points are earned only if a displayed equation contains at least one negative number. One point is given to the team with the qualifying computation closest to zero. Ties give each team a point. Bonus points are received for each negative number used; subtraction signs, however, do not earn points.

   Examples:

   Team X: $^{-}5 - (^{-}4) - (^{-}1) = 0$ Four points are awarded since the equation includes three negative numbers.

   Team Y: $2 + 4 - 6 = 0$, No points are awarded since the equation includes no negative numbers.

   Team X receives an additional point for having the qualifying equation closest to zero.

   Team Y receives no points because it used no negative numbers.
6. Teams play additional rounds and continue to accumulate points.

Making Connections

Promote reflection and make mathematical connections by asking:

- What helped you use negative integers in your equations?

Negatives Score

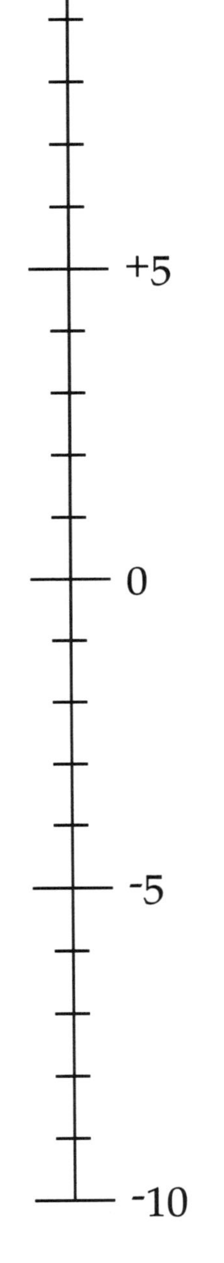

_______________ Team

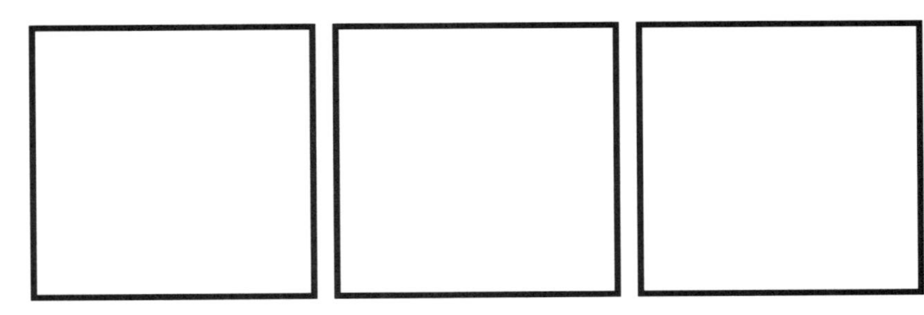

_______________ Team

How?

Topic: Operations with Integers

Object: Use selected integers to form expressions for specific integers.

Groups: Whole class or small group

Tip By allowing use of any number of neighboring integers, students will be able to score many more points.

Materials

- transparency of *How?* activity form, p. 111
- two 5-by-8 cards to mask unwanted integers

Directions

1. The leader displays only one row or column of integers and announces a target number.

Example: The leader says, "Can you make negative 4 by using neighboring numbers and any operations?" With the first row, $^-1$ $^-3$ 5 3 $^-2$ 6 $^-4$, displayed, a student might respond with $(5 + 3) \div {}^-2$, or $^-2 \times 6 \div 3$.

2. Students score one point for each different operation used. Students individually tally points for their expressions.

Example: Two points are scored for $^-2 \times 6 \div 3$

Some other possible target numbers for row 1: 15, 12, 6, 1, 0, $^-5$, and $^-11$

3. Be sure that students use the standard order of operations when writing their expressions.

Examples: $5 + 3 \times {}^-2$ for $^-1$ and $(5 + 3) \times {}^-2$ for $^-16$

-1	-3	5	3	-2	6	-4

4. When students become skilled with the single rows and columns format, the leader may display two rows or a 3-by-3 array. The leader follows the same procedure by announcing target numbers and having the students identify qualifying expressions. If a 3-by-3 array is displayed, students can also use diagonals.

Making Connections

Promote reflection and make mathematical connections by asking:

- Which target numbers are often made more than one way?

How?

$^{-}1$	$^{-}3$	5	3	$^{-}2$	6	$^{-}4$
4	$^{-}6$	$^{-}1$	$^{-}2$	5	$^{-}3$	$^{-}6$
$^{-}5$	2	4	6	$^{-}1$	$^{-}2$	7
$^{-}2$	$^{-}4$	6	1	$^{-}3$	5	$^{-}1$
6	$^{-}3$	$^{-}1$	5	$^{-}2$	$^{-}4$	3
$^{-}3$	4	2	$^{-}5$	$^{-}1$	6	$^{-}2$

Date ____________________ Name ________________________________

Negative Notions 1

STOP Don't start yet! Star problems among 1–4 that may have negative answers.

1. $^{-}7 + 10 =$ _____ **2.** $6 - (^{-}4) =$ _____ **3.** $^{-}15 + 2 - (^{-}3) =$ _____ **4.** $^{-}5 \times 9 =$ _____

Write + or – in each box to complete the equation.

5. □15 + □18 = $^{-}3$ **6.** □12 – □4 = $^{-}8$

7. Ring three neighboring integers whose sum is $^{-}19$. 6 $^{-}21$ 4 $^{-}2$ $^{-}18$

8. Name two consecutive integers whose sum is $^{-}17$. ____________________

Use any three of the integers $^{-}6$, $^{-}4$, 3, and 8, along with any operation signs, to complete the equations.

9. □ ○ □ ○ □ = $^{-}18$ **10.** □ ○ □ ○ □ = $^{-}6$

Go On Use all four of the integers $^{-}6$, $^{-}4$, 3, and 8, along with any operation signs, to complete the equations.

□ ○ □ ○ □ ○ □ = 1

Date ____________________ Name ________________________________

Negative Notions 2

STOP Don't start yet! Star a problem among 1–4 that may have the least answer.

1. $3 + {^{-}8} =$ _____ **2.** $^{-}8 - 6 =$ _____ **3.** $17 - (^{-}3) - 5 =$ _____ **4.** $^{-}36 \div 4 =$ _____

Write + or – in each box to complete the equation.

5. □12 + □5 = 7 **6.** □8 – □5 = $^{-}13$

7. Ring four neighboring integers whose sum is $^{-}7$. 8 $^{-}14$ $^{-}6$ 5 $^{-}10$ 3

8. Name three consecutive integers whose sum is $^{-}18$. ____________________

Use any three of the integers $^{-}3$, 2, 4, and 7, along with any operation signs, to complete the equations.

9. □ ○ □ ○ □ = $^{-}14$ **10.** □ ○ □ ○ □ = $^{-}1$

Go On Use all four of the integers $^{-}3$, 2, 4, and 7, along with any operation signs, to complete the equations.

□ ○ □ ○ □ ○ □ = $^{-}2$ □ ○ □ ○ □ ○ □ = 4

Date ____________ Name ____________

Negative Notions 3

STOP Don't start yet! Star problems among 1–4 that may have positive answers.

1. $^-8 + {}^-7 =$ ____ **2.** $^-4 - (^-8) =$ ____ **3.** $^-13 - 8 + {}^-6 =$ ____ **4.** $^-6 \times {}^-7 =$ ____

Write + or – in each box to complete the equation.

5. ☐ 16 + ☐ 3 = $^-19$ **6.** ☐13 – ☐6 = 19

7. Ring three neighboring integers whose sum is $^-10$. 9 $^-7$ 5 $^-8$ 2

8. Name two consecutive odd integers whose sum is $^-24$. ____________

Use any three of the integers $^-7$, $^-5$, 1, and 4, along with any operation signs, to complete the equations.

9. ☐ ○ ☐ ○ ☐ = 34 **10.** ☐ ○ ☐ ○ ☐ = 2

Go On Use all four of the integers $^-7$, $^-5$, 1, and 4, along with any operation signs, to complete the equations.

☐ ○ ☐ ○ ☐ ○ ☐ = 7 ☐ ○ ☐ ○ ☐ ○ ☐ = $^-3$

Date ____________ Name ____________

Negative Notions 4

STOP Don't start yet! Star problems among 1–4 that may have answers between $^-2$ and 2.

1. $19 + {}^-19 =$ ____ **2.** $9 - (^-5) =$ ____ **3.** $14 + {}^-6 - 7 =$ ____ **4.** $45 \div {}^-9 =$ ____

Write + or – in each box to complete the equation.

5. ☐14 + ☐7 = 7 **6.** ☐ 9 – ☐ 7 = $^-2$

7. Ring four neighboring integers whose sum is 3. $^-8$ 12 $^-9$ 15 $^-7$ 4

8. Name three consecutive even integers whose sum is $^-30$. ____________

Use any three of the integers $^-9$, $^-6$, 2, and 5, along with any operation signs, to complete the equations.

9. ☐ ○ ☐ ○ ☐ = $^-15$ **10.** ☐ ○ ☐ ○ ☐ = 3

Go On Use all four of the integers $^-9$, $^-6$, 2, and 5, along with any operation signs, to complete the equations.

☐ ○ ☐ ○ ☐ ○ ☐ = 13 ☐ ○ ☐ ○ ☐ ○ ☐ = 7

Date _______________ Name _______________________

Negative Notions 5

STOP Don't start yet! Star problems among 1–4 that may have answers between $^{-}6$ and 6.

1. $^{-}9 + 5 =$ _____ **2.** $^{-}2 - 3 =$ _____ **3.** $^{-}12 - (^{-}9) - (^{-}4) =$ _____ **4.** $^{-}8 \times {}^{-}6 =$ _____

Write + or – in each box to complete the equation.

5. □ 17 + □ 5 = 12 **6.** □ 7 – □ 3 = $^{-}4$

7. Ring three neighboring integers whose sum is 7. 11 $^{-}9$ 4 $^{-}3$ 6

8. Name two consecutive even integers whose sum is $^{-}14$. ____________________

Use any three of the integers $^{-}8$, $^{-}5$, 3, and 10, along with any operation signs, to complete the equations.

9. □ ○ □ ○ □ = $^{-}13$ **10.** □ ○ □ ○ □ = $^{-}6$

Go On Use all four of the integers $^{-}8$, $^{-}5$, 3, and 10, along with any operation signs, to complete the equations.

□ ○ □ ○ □ ○ □ = 10

□ ○ □ ○ □ ○ □ = 7

Date _______________ Name _______________________

Negative Notions 6

STOP Don't start yet! Star a problem among 1–4 that may have the greatest answer.

1. $^{-}3 + {}^{-}5 =$ _____ **2.** $^{-}7 + (^{-}7) =$ _____ **3.** $16 - (^{-}7) + {}^{-}5 =$ _____ **4.** $^{-}54 \div {}^{-}6 =$ _____

Write + or – in each box to complete the equation.

5. □ 13 + □ 20 = $^{-}33$ **6.** □ 14 – □ 8 = 22

7. Ring four neighboring integers whose sum is 0. $^{-}6$ $^{-}7$ 13 $^{-}8$ 2 5

8. Name three consecutive odd integers whose sum is $^{-}15$. ____________________

Use any three of the integers $^{-}7$, $^{-}4$, 6, and 9, along with any operation signs, to complete the equations.

9. □ ○ □ ○ □ = 12 **10.** □ ○ □ ○ □ = 4

Go On Use all four of the integers $^{-}7$, $^{-}4$, 6, and 9, along with any operation signs, to complete the equations.

□ ○ □ ○ □ ○ □ = 8 □ ○ □ ○ □ ○ □ = $^{-}6$

Integer Four-in-a-Row

Topic: Subtracting Integers

Object: Cover four numbers in a row.

Groups: 2 players

Materials for each group

- *Integer Four-in-a-Row* gameboard, p. 116
- 2 transparent chips
- different kind of markers for each player

Tip *Introduce this game with a transparency of the gameboard and half the class playing against the other half.*

Directions

1. The first player places two transparent chips at the bottom of the gameboard, indicating two integers. The same player finds the difference of the two selected integers and places a marker on the resulting difference.
2. The other player moves only one of the transparent chips to a new integer. Next, this player subtracts either integer from the other and places a marker on that difference.
3. Players continue alternating turns, moving one transparent chip each time, subtracting the integers and covering the difference on the gameboard.
4. The winner is the first player to have four markers in a row horizontally, vertically, or diagonally.

-6	1	10	-7	-5	4
-10	-3	-12	1	11	-10
-9	2	7	6	-1	2
7	9	-2	-8	-2	9
-1	10	-11	-9	3	-7
5	-2	8	-1	12	-4

2 4 5 6

-3 -4 -5 -6

Making Connections

Promote reflection and make mathematical connections by asking:

- What made it easier to cover adjacent numbers in this variation of a familiar game?
- What strategies helped you line up your markers in a row?

Integer Four-in-a-Row

-6	1	0	-7	-5	4
-10	-3	-12	1	11	-10
-9	2	7	6	-1	2
7	9	-2	-8	-2	9
-1	10	-11	-9	3	-7
5	-2	8	-1	12	-4

2 4 5 6

-3 -4 -5 -6

Cover It!

Topic: Operations with Integers

Object: Cover every displayed integer.

Groups: 2 players or pair players

Tip Encourage students to create variations of this game to provide additional practice with negative integers.

Materials per group

- *Cover It!* gameboard, p. 118
- 6 Number Cubes (1-6)
- 11 markers for each player or pair players

Directions

1. Each player rolls three Number Cubes to generate three integers. The player must use each generated number. Each player is allowed to use any operation and sign with the three numbers to produce an expression for an uncovered integer.

   Example: If 1, 2, and 4 are rolled, the player might think $^{-}4 \div 2 + 1$ and cover $^{-}1$; or think $^{-}1 + 2 + {}^{-}4$ and cover $^{-}3$; or think $4 \div {}^{-}2 \times {}^{-}1$ and cover 2.
2. Each player informs the opposing player how the three rolled numbers were used to yield the uncovered integer.
3. Play continues with players rolling Number Cubes, creating appropriate expressions, and covering resulting integers. If a player is unable to create a computation for an uncovered integer, no integer is covered for that turn.
4. The first player to cover all eleven of his or her numbers wins. It's possible for both players to win if they both complete covering the integers in the same number of turns.

Making Connections

Promote reflection and make mathematical connections by asking:

- Which numbers seemed easier to cover? How can this be explained?
- Which rolled combinations seemed to generate more possibilities? Please explain your reasoning.

Cover It!

-5	5
-4	4
-3	3
-2	2
-1	1
0	0
1	-1
2	-2
3	-3
4	-4
5	-5

Neighboring Integers Count

Topic: Operations with Integers

Object: Score the greatest total.

Groups: 2 to 4 players

Tips *To avoid wait time, allow each player to draw three squares and simultaneously work on solutions. To provide a challenge, have students draw and use four squares.*

Materials for each group

- *Neighboring Integers Count* gameboard, p. 120
- set of Negative Integer Squares, p. 148
- set of Digit Squares (0 removed), p. 147
- 10–18 markers for each player
- opaque container

Directions

1. The Negative Integer Squares and Digit Squares are mixed and placed in the container.

2. The first player draws three Squares and uses any operations to make an expression for a negative integer from the three numbers. After stating the expression, the player locates the resulting answer and covers it with a marker.

Example: With 1, $^{-}2$, and $^{-}5$, a player might compute $^{-}5 - {}^{-}2 + 1$ and cover $^{-}2$ or compute $(^{-}5 - {}^{-}2) \times 1$ and cover $^{-}3$.

-18	-17	-16	-15	-14	-13
-12	-11	-10	-9	-8	-7
-6	-5	-4	-3	-2	-1
1	2	3	4	5	6
7	8	9	10	11	12
13	14	15	16	17	18

3. The next player draws three Squares, forms an expression, and computes a new answer to equal an uncovered number on the board. One point is scored if the player covers a number that shares a side with any already-covered number.

4. Players record any score earned after each turn. Sometimes a player will be unable to cover any available numbers and must pass for that turn. The game ends after all numbers are covered, or after three consecutive passes by one player.

5. The player with the greatest point total wins.

Making Connections

Promote reflection and make mathematical connections by asking:

- What kinds of draws do you prefer? Please explain.
- What strategies worked to give you high totals on your turns?

Neighboring Integers Count

-18	-17	-16	-15	-14	-13
-12	-11	-10	-9	-8	-7
-6	-5	-4	-3	-2	-1
1	2	3	4	5	6
7	8	9	10	11	12
13	14	15	16	17	18

Date ______________ Name ______________________

Finding Sums and Differences

Ring four neighboring integers that yield each sum shown.

1. Sum = 2	2. Sum = -10	3. Sum = -2	4. Sum = -12	5. Sum = -9	6. Sum = 1
-6	7	6	9	5	-13
13	-12	-4	-15	-17	17
-5	9	8	8	14	13
-4	-15	-5	-14	-12	-16
3	8	-9	6	6	8
8	8	4	9	-15	-6

Write + or – in each box to complete the equation.

7. □13 – □9 = -4 **8.** □16 – □8 = -24 **9.** □12 – □7 = 19

10. □14 – □6 = 20 **11.** □8 – □5 = -3 **12.** □13 – □7 = -20

Find a trail through the integers by subtracting from the START number to reach the STOP number. You may move vertically or diagonally.

13.

Start 7	
5	-8
-6	7
1	-4
12 Stop	

14.

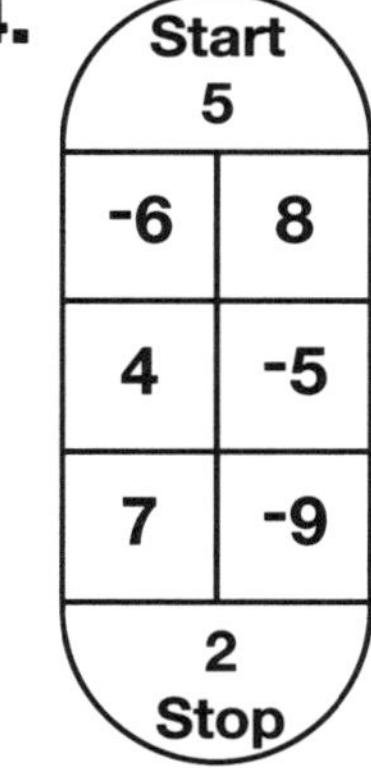

15.

Start 4	
-7	3
-2	6
-5	9
-6 Stop	

16.

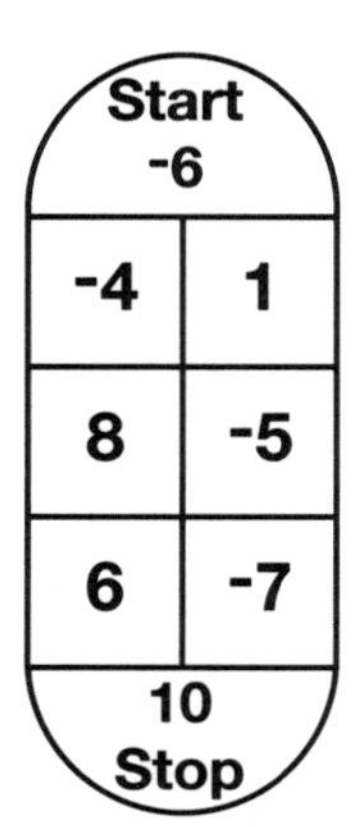

17.

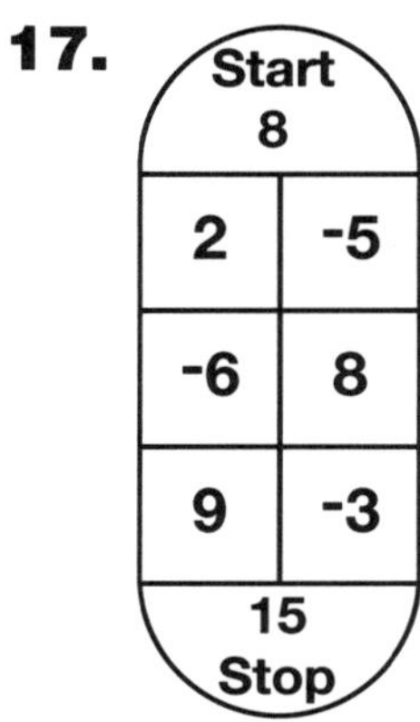

18.

Date ____________ Name ____________

Finding Tic-Tac-Toes I

Complete each equation. Mark the tic-tac-toe row that contains three of the answers from the six computations at the right.

1.

-2	3	-9
-11	-7	14
-1	-3	22

$3 + {}^{-}5 =$ ______ $\quad {}^{-}4 - 5 =$ ______

${}^{-}4 + {}^{-}7 =$ ______ $\quad 6 - 13 =$ ______

${}^{-}8 + 5 =$ ______ $\quad 18 - {}^{-}4 =$ ______

2.

-11	9	-2
-21	-8	-16
-14	31	11

$16 + {}^{-}27 =$ ______ $\quad 21 - {}^{-}10 =$ ______

${}^{-}6 + 15 =$ ______ $\quad 7 - 9 =$ ______

${}^{-}8 + {}^{-}6 =$ ______ $\quad 15 - 23 =$ ______

3.

35	-28	-31
-13	16	-9
-35	-5	-21

${}^{-}46 + 18 =$ ______ $\quad {}^{-}18 - 13 =$ ______

$30 - 43 =$ ______ $\quad {}^{-}17 + 33 =$ ______

${}^{-}15 + {}^{-}6 =$ ______ $\quad {}^{-}20 - 15 =$ ______

4.

19	13	46
-43	7	-16
36	-5	-13

${}^{-}14 + 50 =$ ______ $\quad {}^{-}12 + {}^{-}31 =$ ______

$62 - 75 =$ ______ $\quad 31 + {}^{-}18 =$ ______

${}^{-}83 + 129 =$ ______ $\quad 7 + {}^{-}12 =$ ______

Trivia The sum of the numbers in the tic-tac-toe row for Problem 4 is the number of hours a three-toed sloth sleeps each day.

Challenge Create a tic-tac-toe problem for your classmates to solve. Be sure your puzzle has only one row that contains three of the answers.

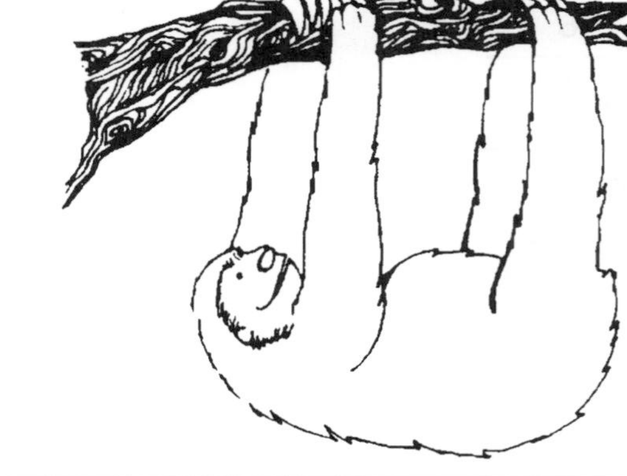

Date ____________ Name ____________

Finding Tic-Tac-Toes II

Complete each equation. Mark the tic-tac-toe row that contains three of the answers from the six computations at the right.

1.

-60	-42	-27
-18	27	-48
66	18	60

$6 \cdot {}^-8 =$ ______

$^-3 \cdot {}^-9 =$ ______

$^-7 \cdot 6 =$ ______

$(5 \cdot {}^-3) \cdot {}^-4 =$ ______

$^-2 \cdot (11 \cdot {}^-3) =$ ______

$({}^-36 \div 4) \cdot 2 =$ ______

2.

-42	36	-4
60	4	24
-28	-36	28

$^-5 \cdot {}^-12 =$ ______

$9 \cdot {}^-4 =$ ______

$^-28 \div {}^-7 =$ ______

$(63 \div {}^-9) \cdot {}^-4 =$ ______

$^-48 \div ({}^-2 \cdot {}^-6) =$ ______

$(42 \div {}^-7) \cdot 7 =$ ______

3.

-56	12	-25
63	-32	4
-63	-12	-64

$7 \cdot {}^-8 =$ ______

$^-50 \div 2 =$ ______

$(36 \div {}^-4) \cdot {}^-7 =$ ______

$60 \div ({}^-30 \div {}^-2) =$ ______

$(8 \cdot {}^-6) \div 4 =$ ______

$^-4 \cdot (48 \div 3) =$ ______

4.

-72	24	30
-30	-3	-20
-12	108	72

$^-45 \div 15 =$ ______

$({}^-8 \cdot {}^-3) \cdot {}^-3 =$ ______

$9 \cdot ({}^-36 \div {}^-3) =$ ______

$80 \div ({}^-36 \div 9) =$ ______

$(8 \cdot {}^-6) \div 4 =$ ______

$({}^-15 \cdot 4) \div {}^-2 =$ ______

Trivia The sum of the numbers in the tic-tac-toe row for Problem 4 is the number of inches across the eye of a giant squid.

Challenge Create a tic-tac-toe problem for your classmates to solve. Be sure your puzzle has only one row that contains three of the answers.

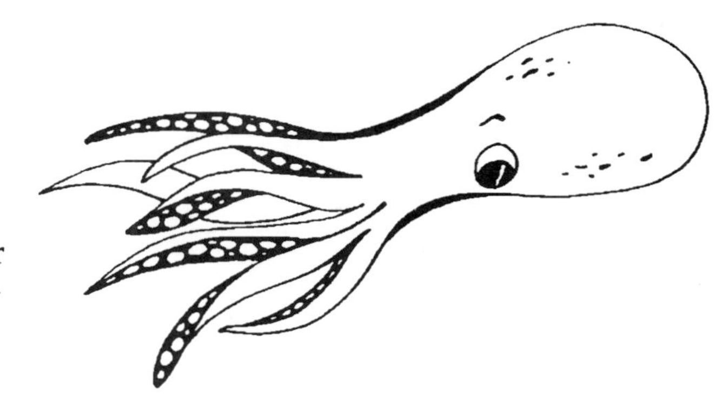

Date ______________ Name ______________________

Integer Choices

Choose and record 4 digits. Digit Choices ____ ____ ____ ____

In each equation, use any three digits as negative or positive intergers. Be sure to include negative signs. Complete each equation using any operations with the standard order of operation, inserting parentheses when necessary.

□ ○ □ ○ □ = 7

□ ○ □ ○ □ = 6

□ ○ □ ○ □ = 5

□ ○ □ ○ □ = 4

□ ○ □ ○ □ = 3

□ ○ □ ○ □ = 2

□ ○ □ ○ □ = 1

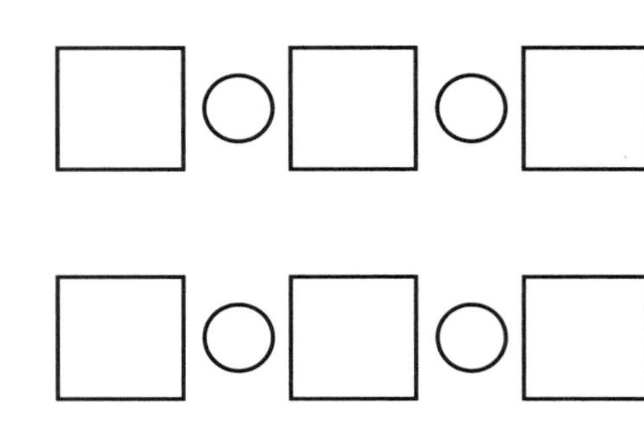

□ ○ □ ○ □ = 0

□ ○ □ ○ □ = -1

□ ○ □ ○ □ = -2

□ ○ □ ○ □ = -3

□ ○ □ ○ □ = -4

□ ○ □ ○ □ = -5

□ ○ □ ○ □ = -6

□ ○ □ ○ □ = -7

Algebra

Assumptions Algebraic concepts and equations have been taught and reviewed, emphasizing understanding and enhancing number sense. Two-colored counters and other visual models, such as algebra tiles, have been used extensively.

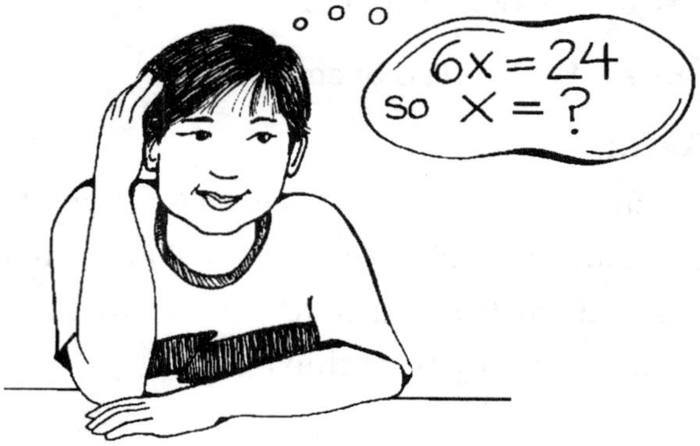

Section Overview and Suggestions

Sponges

Modeling Equations pp. 126–128

Phrases to Symbols p. 129

What's My Rule? pp. 130–131

Creating Equations p. 132

These repeatable, whole-class or small-group warm-ups build algebra sense as students demonstrate their abilities to model algebraic expressions and enhance their abilities to solve algebraic equations.

Skill Checks

Equation Station 1–6 pp. 133–135

The Skill Checks provide a way for parents, students, and you to see students' improvement with algebra concepts. Copies can be cut in half so Checks may be used at different times. Remember to have all students respond to STOP, the number sense task, before solving the ten problems.

Games

Finding Unknowns pp. 136–137

Create and Solve pp. 138–139

Solve and Travel pp. 140–142

These open-ended and repeatable Games actively involve students in solving algebraic equations while enhancing their strategic thinking abilities. Adequate use of the *Creating Equations* Sponge should ensure student success with *Create and Solve*, a Game that requires students to generate qualifying equations. With frequent use of these Games at home or in family math sessions, students will gain confidence in their abilities to engage in more abstract levels of mathematics.

Independent Activities

Symbolizing Number Riddles p. 143

Geographic Discoveries via Algebra p. 144

Each Independent Activity engages students in practice of writing and solving algebraic equations. Both *Symbolizing Number Riddles* and *Geographic Discoveries via Algebra* encourage students to create similar riddles and puzzles for classmates to solve. The use of these student-authored activities provides high-interest additional practice of challenging and important math skills.

Modeling Equations

Tip *Allow groups who use a different approach to describe and demonstrate their solution process.*

Topic: Representing Algebraic Equations Visually and Concretely

Object: Use objects to represent a displayed equation.

Groups: Whole class or small group

Materials

- transparency of *Modeling Equations* activity form, p. 127
- transparency of *Equation Strips,* cut apart, p. 128
- 12 objects such as tiles to represent variables
- 2 colors of transparent chips

Directions

1. The leader places Equation Strips in a container and asks a volunteer to draw one and display the equation on the activity form.
2. After students confer with a partner, one student volunteers to represent the displayed equation visually, using objects and colored chips. Variables should be represented by the objects, while positive integers should be designated by one color of chips and negative integers by the other color.

Example: $3x + 1 = 10$

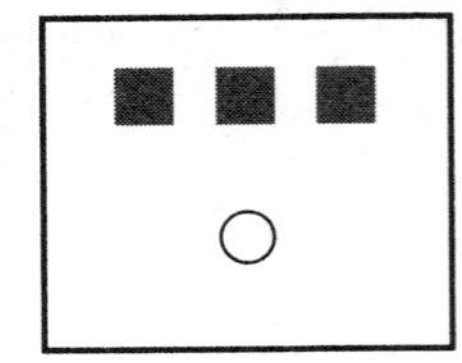

=

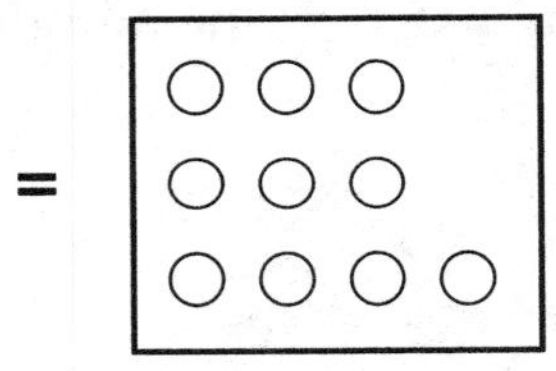

3. Students pair up to discuss how to use displayed materials to solve the equation. One pair volunteers to describe and demonstrate a solution to the entire class.
4. Another student volunteers to record the pair's solution steps symbolically.
5. Once a solution to an equation is satisfactorily modeled, a new equation is displayed and the previous steps are repeated.

Making Connections

Promote reflection and make mathematical connections by asking:

- Which equations are more difficult to represent? Explain.

Modeling Equations

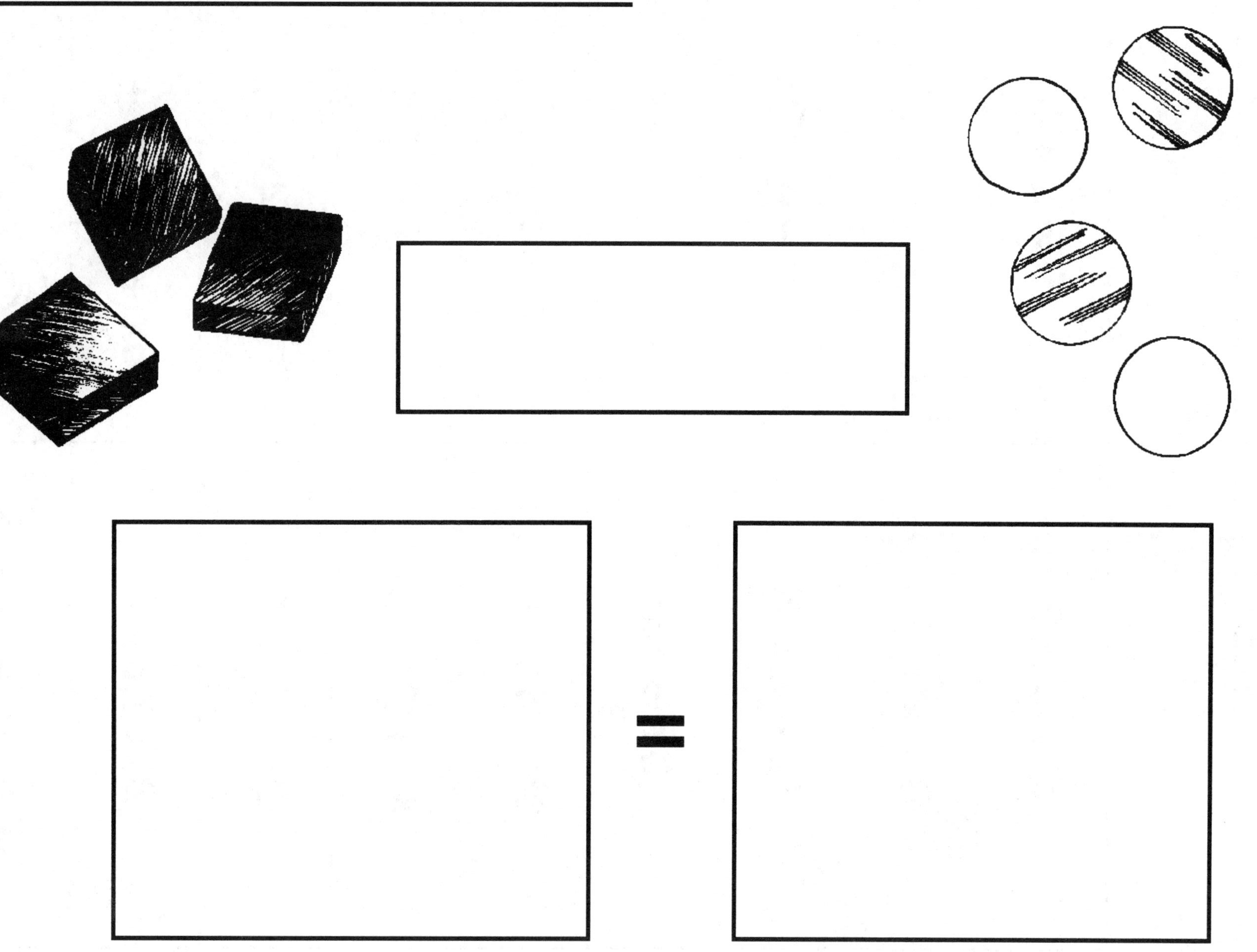

Equation Strips

$x + 3 = 8$	$9 = x + 5$
$9 + x = 15$	$7 = x - 2$
$12 - x = 5$	$x - 4 = 7$
$6x = 18$	$x \cdot 3 = 15$
$3x + 1 = 10$	$3 + 2x = 13$
$4x + 2 = 10$	$\frac{x}{2} = 7$
$3x - 3 = 9$	$14 - 4x = 2$
$8 = 7x - 6$	$6x + 2 = 4x + 10$
$5x + 16 = 7x + 4$	$3x - 7 = 9 - 5x$

Phrases to Symbols

Topic: Algebraic Representations

Object: Translate word phrases into algebraic expressions.

Groups: Whole class or small group

Materials

- teacher prepared word phrases

Directions

1. The leader shares a word phrase.
   Example: "four more than an unknown number"
2. The leader asks students to translate the phrase into symbols. Students then work in pairs to compare their solutions.
3. One or more pairs may volunteer to share their expressions (in this case, $x + 4$).
4. After a word phrase is correctly represented algebraically, a new phrase is given and the process is repeated.

   Possible phrases:

   an unknown number divided by four ($x \div 4$)
   seven less than an unknown amount ($x - 7$)
   the quotient of some number divided by five ($\frac{x}{5}$)
   three more than five times a number ($5x + 3$)
   the product of some number and six ($6x$)
   three more than four times a number ($4x + 3$)
   one third of a number, decreased by two ($\frac{1}{3}x - 2$)
   the product of a number and five, increased by eight ($5x + 8$)

Tips *Have students work in pairs to prepare and present their own word phrases for classmates to translate. A more challenging level would be to present algebraic expressions and have students translate them into phrases. Encourage variety in translations.*

Making Connections

Promote reflection and make mathematical connections by asking:

- What helped you get started in translating a phrase into symbols?
- Which phrases were more difficult to represent algebraically? Explain.

What's My Rule?

Tip *Superimpose a blank transparency on the activity form to easily reuse the T-table for future rounds.*

Topic: Finding Algebraic Rules

Object: Identify an algebraic function.

Groups: Whole class or small group

Materials

- transparency of *What's My Rule?* activity form, p. 131
- prepared rules for beginning rounds

Directions

1. The leader displays the *What's My Rule?* activity form and provides one example of input and output based on a prepared but hidden rule. This action is recorded in the "In/Out" T-table.

   Example: "If 3 is the input number, the output number is 11."

2. A student volunteers another input number, and the leader identifies the resulting output number, which is also recorded on the same T-table.
3. As students continue to suggest new input numbers, other students try to identify the corresponding output numbers.
4. After a majority of students are able to identify the output numbers for given input numbers, students are asked to state the rule in an algebraic form.
5. The leader begins a new round with another rule and beginning example, following the previous four steps. Gradually, more complex equations are used.

   Possible rules: $4x + 1$ $\frac{3x}{2}$ $2x - 5$ $7 - 3x$

6. If students work in pairs to create new rules and beginning examples, this becomes a student-led activity.

Making Connections

Promote reflection and make mathematical connections by asking:

- What helped you translate the identified input/output pattern into an algebraic expression?
- Why is it necessary to have at least two examples before identifying possible rules?

What's My Rule?

In	Out

Rule

Creating Equations

Topic: Creating Equations for Unknowns

Object: Create as many qualifying equations as possible.

Groups: Whole class or small group

Tips *Initially have students work as pairs to increase success rate. For a greater challenge, follow the same procedures but have individual students try to generate five different equations for each of the variable values 2 through 6.*

Materials

- 3 Number Cubes (1–6), p. 149
- blank paper

Directions

1. Have each student draw three boxes at the top of a blank sheet of paper.
2. Three Number Cubes are rolled to generate three numbers that are recorded in the students' three boxes.
3. Students are to use the three displayed numbers to create an equation containing a variable. Each student selects a value from 2 through 6 for the value of the variable. If the chosen value for the variable works in an equation, the equation is recorded as a qualifying equation.
4. Students attempt to create a variety of equations that have 2 through 6 as solutions.

   Example: 2, 3, and 5 are rolled. If $x = 4$, then a qualifying equation is $2x - 3 = 5$. If $x = 2$, then $3x = 5 + \frac{2}{x}$; if $x = 6$, then $x - 5 = 3 - 2$; if $x = 4$, then $2x - 3 = 5$.

2  

$2x - 3 = 5$

$x = 4$

5. Different students share their equations for the five possible values (2, 3, 4, 5, and 6) of the variable.
6. If time allows, three new numbers are generated and the steps are repeated.

Making Connections

Promote reflection and make mathematical connections by asking:

- What helped you create qualifying equations?
- What strategy did you use to find multiple equations?

Date ____________ Name ____________

Equation Station 1

STOP Don't start yet! Star problems that may have answers between 20 and 40.

Give the value of each expression for $n = 3$.

1. $6n + 14$ ______ **2.** $20 - 3n$ ______

Solve each equation for x.

3. $13 = x + 8$

4. $3x = 21$

5. $\frac{x}{4} = 9$

6. $2.8x = 0.84$

7. $6x + 25 = 49$

8. $4(x - 8) = 52$

9. $4x - 6 + 2x = 24$

10. $\frac{2(6x - 10)}{4} = 7$

Go On Complete the table.

x	y	$x + y$	xy
20		52	
	18		396

Date ____________ Name ____________

Equation Station 2

STOP Don't start yet! Star problems that may have answers less than 10.

Give the value of each expression for $t = 7$.

1. $8t - 12$ ______ **2.** $15 + 4t$ ______

Solve each equation for x.

3. $x - 4 = 11$

4. $112 = 8x$

5. $12 = \frac{x}{6}$

6. $3x = 7.2$

7. $9x - 15 = 174$

8. $2(x + 4) = 24$

9. $8x + 6x = {}^{-}42$

10. $\frac{6(3x - 1)}{8} = 6$

Go On Choose any number. Multiply it by 3. Add 12. Divide by 3. Subtract your original number. Try other original numbers. Why is the answer always the same? Write your own riddle.

Date ____________ Name ____________________

Equation Station 3

STOP Don't start yet! Star problems that may have answers greater than 50.

Give the value of each expression for $y = 4$.

1. $11y + 15$ _______ **2.** $42 - 4y$ _______

Solve each equation for x.

3. $x + 27 = 58$ **4.** $6x = 54$ **5.** $15 = \frac{x}{3}$

6. $0.2x = 34$ **7.** $8x - 10 = 62$ **8.** $76 = 4(x + 9)$

9. $12x - 7x = 50$ **10.** $\frac{3(4x + 6)}{6} = 9$

Go On Complete the table.

B	C	$B - C$	$B \div C$
512			16
		60	4

Date ____________ Name ____________________

Equation Station 4

STOP Don't start yet! Star problems that may have answers less than 10.

Give the value of each expression for $b = 20$.

1. $8b - 23$ _______ **2.** $42 + 4b$ _______

Solve each equation for x .

3. $19 = x - 52$ **4.** $11x = 132$ **5.** $\frac{x}{7} = 9$

6. $0.04x = 0.6$ **7.** $15x + 5 = 230$ **8.** $7(x + 2) = 56$

9. $6x + 2 + 4x = 42$ **10.** $\frac{4(7x - 21)}{7} = 8$

Go On Choose any number. Add 5. Multiply by 2. Subtract 10. Divide by 2. What is your answer? Try other original numbers. What do you notice? Try to explain. Write your own riddle.

Date ______________ Name ______________________

Equation Station 5

STOP Don't start yet! Star a problem that may have the greatest answer.

Give the value of each expression for $a = 5$.

1. $23 + 7a$ _______ **2.** $11a - 60$ _______

Solve each equation for x.

3. $8 = x + 23$ **4.** $7x = 56$ **5.** $15 = \frac{x}{5}$

6. $3.2x = 0.96$ **7.** $7x + 120 = 204$ **8.** $4(2x - 7) = 68$

9. $4x + 8x = 48$ **10.** $\frac{3(5x + 2)}{4} = 9$

Go On Complete the table.

M	N	$M \times N$	$M - N$
	37		18
	12	600	

Date ______________ Name ______________________

Equation Station 6

STOP Don't start yet! Star problems that may have answers less than 5.

Give the value of each expression for $c = 8$.

1. $8c - 50$ _______ **2.** $4c + 90$ _______

Solve each equation for x.

3. $x - 40 = 12$ **4.** $90 = 15x$ **5.** $\frac{x}{8} = 11$

6. $0.15x = 1.05$ **7.** $5x - 57 = {}^{-}17$ **8.** $8(7x - 4) = 80$

9. ${}^{-}19 = 6x - 5 + x$ **10.** $\frac{2(7x + 13)}{3} = 18$

Go On Choose any number. Multiply it by 5. Add 15. Divide by 5. Subtract your original number. Try other original numbers. Why is the answer always the same? Write your own riddle.

Finding Unknowns

Topic: Solving Equations

Object: Cover four numbers in a row.

Groups: 2 players

Tip *Encourage students to design new gameboards for additional practice with their classmates.*

Materials for each group

- *Finding Unknowns* gameboard, p. 137
- 2 transparent chips
- different kind of markers for each player

Directions

1. The first player places one transparent chip in each of the two rows below the gameboard, indicating two sides of an equation. The same player finds the value of x and places a marker on the corresponding square in the grid.
2. The other player moves only one of the transparent chips to a new amount. This player then finds the value of x and places a marker on that corresponding square in the grid.
3. Players alternate turns, moving one transparent chip each time, solving for x, and covering the corresponding square in the grid.
4. When a player creates an expression that does not equal any uncovered number on the gameboard, that player loses a turn.
5. The winner is the first player to have four markers in a row horizontally, vertically, or diagonally.

3	6	4	2	5	4
4	5	12	6	8	3
9	2	5	4	3	6
6	10	3	8	2	4
2	4	9	6	5	3

$3x$ $4x$ $(5x)$ $6x$ $8x$ $9x$

= 12 (15) 16 18 20 24 30 36

Making Connections

Promote reflection and make mathematical connections by asking:

- What are some good beginning plays? Explain.
- What strategies helped you get your markers in a row?

Finding Unknowns

3	6	4	2	5	4
4	5	12	6	8	3
9	2	5	4	3	6
6	10	3	8	2	4
2	4	9	6	5	3

3*x* 4*x* 5*x* 6*x* 8*x* 9*x*

= 12 15 16 18 20 24 30 36

Create and Solve

Topic: Writing and Solving Equations

Object: Cover each number on a number strip.

Groups: Pair players or 2 players

Tip *To challenge your more able students, you can significantly increase the difficulty of this game by changing the positive integers on the gameboard into negative integers and allowing any signs for the generated numbers.*

Materials for each group

- *Create and Solve* gameboard, p. 139
- 6 markers for each player
- 3 Number Cubes (1–6), p. 149

Directions

1. A player from one pair rolls the Number Cubes and displays them in the boxes on the gameboard.
2. Each pair uses the three generated numbers and a variable to create an expression whose value is a number on its number strip.
3. Once a pair decides on an expression, a corresponding equation is recorded on another sheet of paper.

   Example: Using 3, 4, and 6, one pair chooses 4 from its number strip and records $\frac{6-x}{4} + 3 = 4$, while the other pair chooses 2 from its number strip and records $\frac{4\,(6)}{3x} = 2$

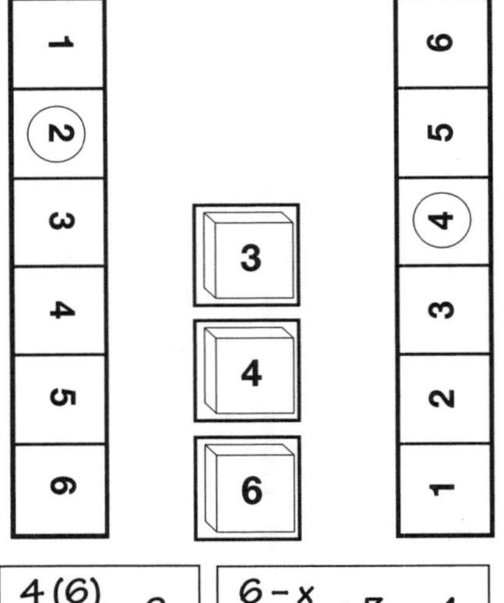

4. Then the papers are exchanged and each pair solves the other pair's equation. Then both pairs look at both solutions.
5. If the pairs agree on a solution, the pair that created the equation is allowed to cover its chosen number, not the value of the unknown, on the number strip. (In the above example, one pair covers 2 while the other pair covers 4.)
6. A player from the second pair rolls the Number Cubes and the process is repeated.
7. The game continues until one pair covers all six of its cells. It's possible that both pairs might win on the same round.

Making Connections

Promote reflection and make mathematical connections by asking:

- Which numbers were more difficult to cover? Try to explain why.

Create and Solve

$$\frac{6-x}{4}+3=4$$

1
2
3
4
5
6

6
5
4
3
2
1

Solve and Travel

Topic: Solving Equations

Object: Create a pathway across a gameboard.

Groups: 2 players

Tip *When students are ready for a greater challenge, have them use* Solve and Travel B *gameboard, p. 142. Note that a* different *special Number Cube (5-5-6-6-7-7), p. 149, will have to be created.*

Materials for each group

- different kind of markers for each player
- special Number Cube (3-3-4-4-5-5), p. 149
- *Solve and Travel A* gameboard, p. 141

Directions

1. After tossing the special Number Cube, the first player looks for cells containing an equation whose solution is the tossed digit. The player gives the solution to the equation and provides an explanation. Then the qualifying cell is covered with that player's marker.

Example: The player rolls a 3 and states, "If $9x - 6 = 21$, $9x = 27$, so $x = 3$."

2. The other player follows the same procedure. Only one marker can occupy a single cell.

3. Players continue alternating turns until one player forms a continuous pathway across the gameboard from left to right or top to bottom.

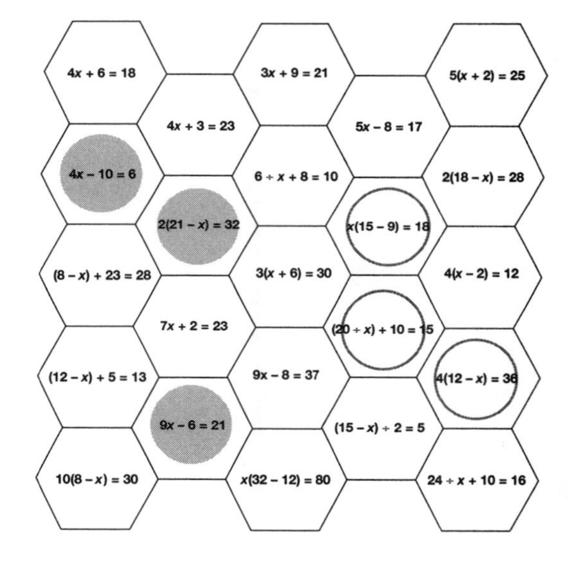

Making Connections

Promote reflection and make mathematical connections by asking:

- What helped you to determine the value for a variable?
- What strategy helped you choose equations in a continuous path?

Solve and Travel A

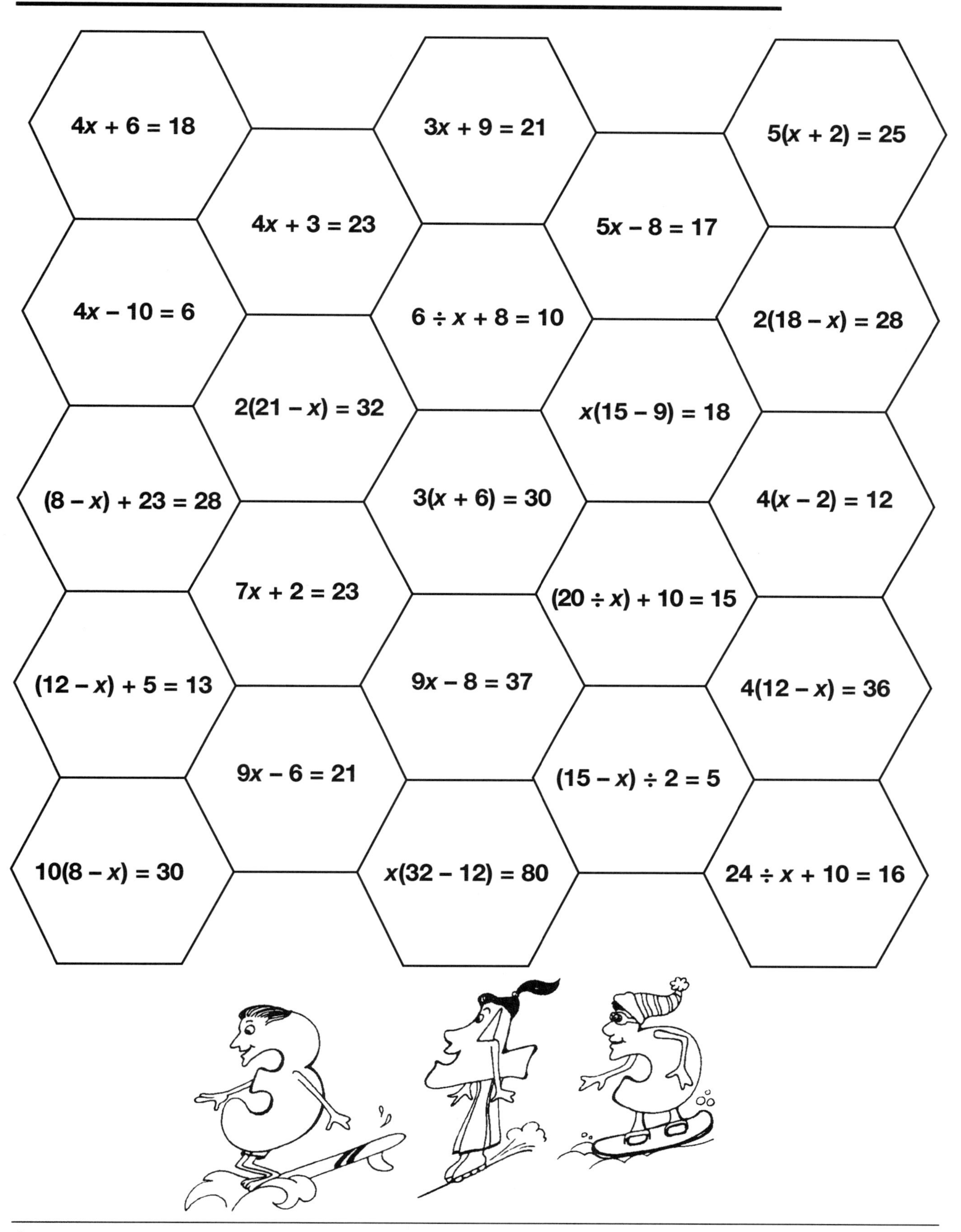

Solve and Travel B

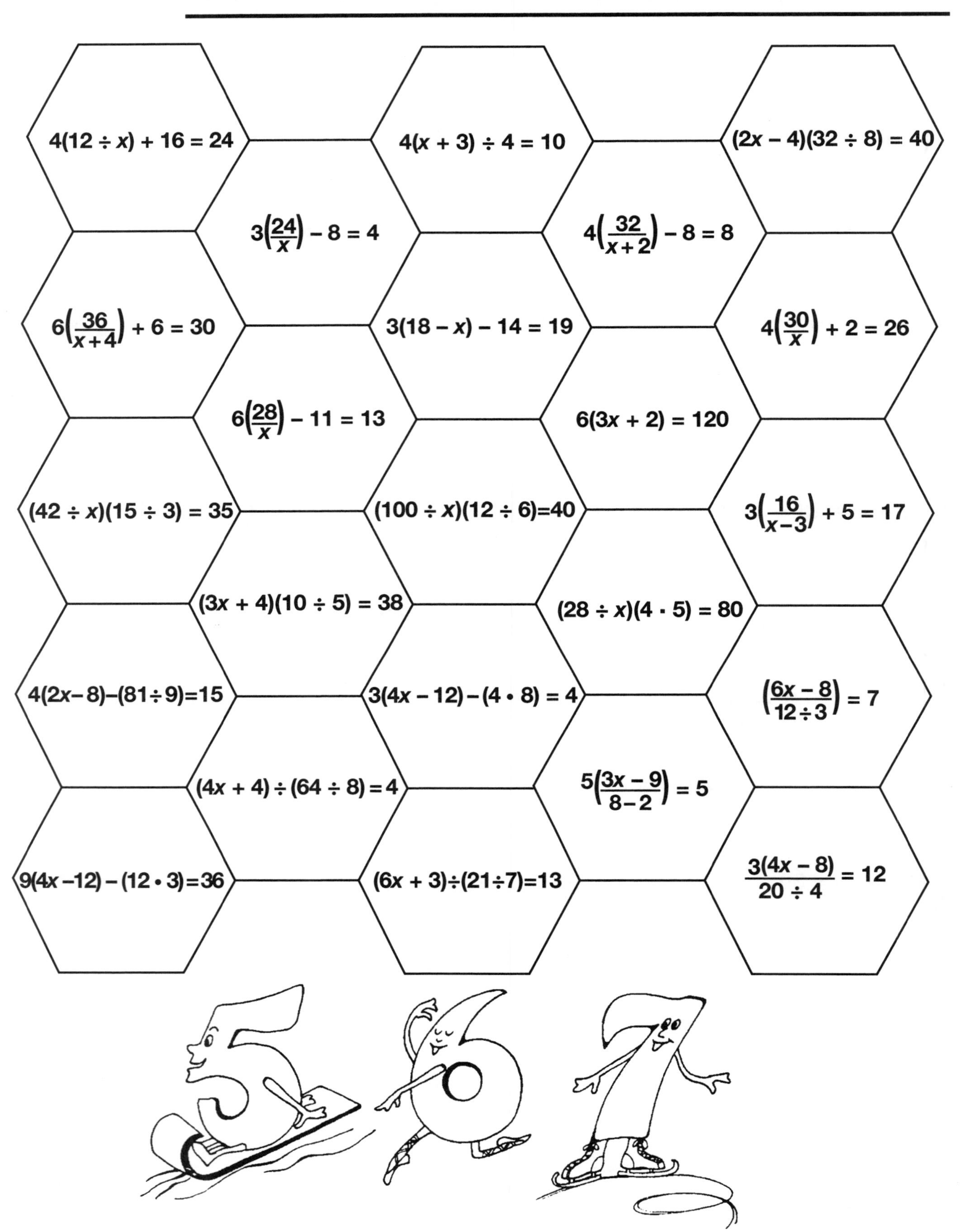

Date ____________ Name ____________________

Symbolizing Number Riddles

Write an equation for each riddle. Then find the number. The first problem is done for you.

1. If you add 17 to the number, you get 25.

   $x + 17 = 25$ ____ $x = 8$

2. If you subtract 9 from the number, you get 15.

   ____________ ______

3. If you have 8 less than the number, you have 30.

   ____________ ______

4. If you have 10 more than the number, you have 51.

   ____________ ______

5. If you triple the number, you get 39.

   ____________ ______

6. If you divide the number by 4, you get 20.

   ____________ ______

7. If you halve the number, you get 36.

   ____________ ______

8. If you multiply the number by 6, you get 54.

   ____________ ______

9. If you add 6 to twice the number, you get 20.

   ____________ ______

10. If you have 4 less than 5 times the number, you have 46.

    ____________ ______

11. If you subtract 15 from 8 times the number, you get 73.

    ____________ ______

12. If you have 9 more than three times the number, you have 60.

    ____________ ______

13. If you have 10 more than $\frac{1}{3}$ of the number, you have 21.

    ____________ ______

14. If you add 28 to $\frac{1}{4}$ of the number, you get 40.

    ____________ ______

15. If you subtract 7 from $\frac{1}{5}$ of the number, you get 9.

    ____________ ______

16. If you have 17 less than half the number, you have 13.

    ____________ ______

17. If you increase the number by 9 and then divide by 5, you get 3.

    ____________ ______

18. If you decrease the number by 15 and then divide by 3, you get 4.

    ____________ ______

Challenge Write a number riddle for a friend to solve.

Date ____________ Name ____________

Geographic Discoveries via Algebra

Solve each equation.

1. $0.08x = 0.24$ **2.** $8x - 7 = 9$ **3.** $\frac{42}{x} = 6$ **4.** $5.8 = 2.3x + 1.2$

5. $12x + 2 = 50$ **6.** $12x - 3 = 8x + 5$ **7.** $4x - 5 = 19$ **8.** $x - 3.7 = 3.3$

9. $5x + 3 = 6x - 1$ **10.** $75x = 600$ **11.** $6x - 5 = 5x + 2$ **12.** $\frac{x + 7}{3} = 4$

13. $0.7x = 6.3$ **14.** $\frac{18}{x} + 14 = 17$ **15.** $4(3x + 2) = 32$

Use the following key with your solutions to the equations to identify the two countries in the world with the longest coastlines. The numbers under the blanks correspond to the exercise numbers above.

2 = A 3 = C 4 = D 5 = E 6 = I 7 = N 8 = O 9 = S

16. ______ ______ ______ ______ ______ ______
1 2 3 4 5 6

This country has 151,485 miles, or 243,791 kilometers, of coastline.

17. ______ ______ ______ ______ ______ ______ ______ ______ ______
7 8 9 10 11 12 13 14 15

This country has 33,999 miles, or 54,716 kilometers, of coastline.

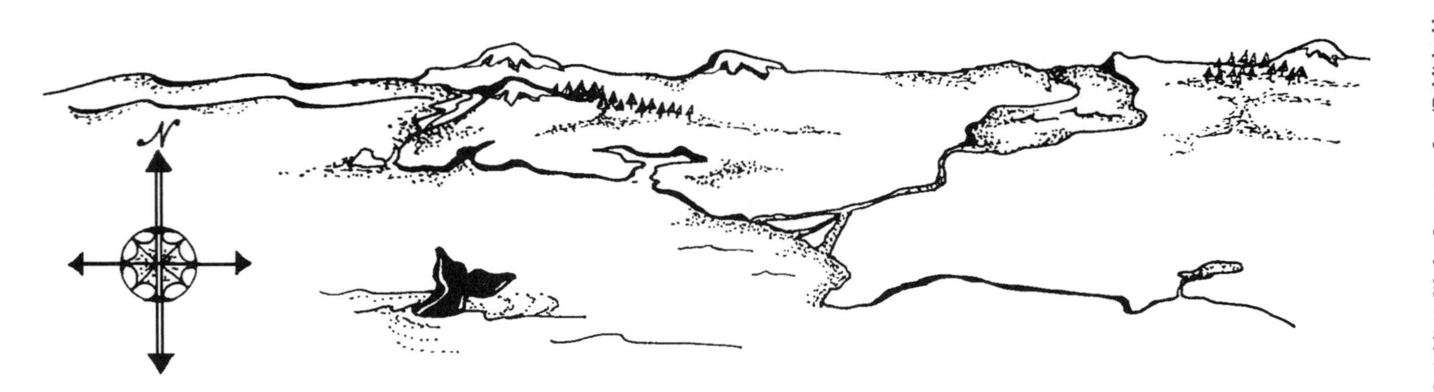

Challenge Create an algebra geography puzzle for your classmates to solve.

Blackline Masters

Digit Cards

Digit Squares

Negative Integer Squares

Number Cubes (1–6, blank)

Spinners (1–6, blank)

Digit Cards

0		
1	2	3
4	5	6
7	8	9

Digit Squares

0	1	2	3	4
5	6	7	8	9

0	1	2	3	4
5	6	7	8	9

0	1	2	3	4
5	6	7	8	9

Negative Integer Squares

	-1	-2	-3	-4
-5	-6	-7	-8	-9

	-1	-2	-3	-4
-5	-6	-7	-8	-9

	-1	-2	-3	-4
-5	-6	-7	-8	-9

Number Cubes

Cut solid lines. Fold on dotted lines.

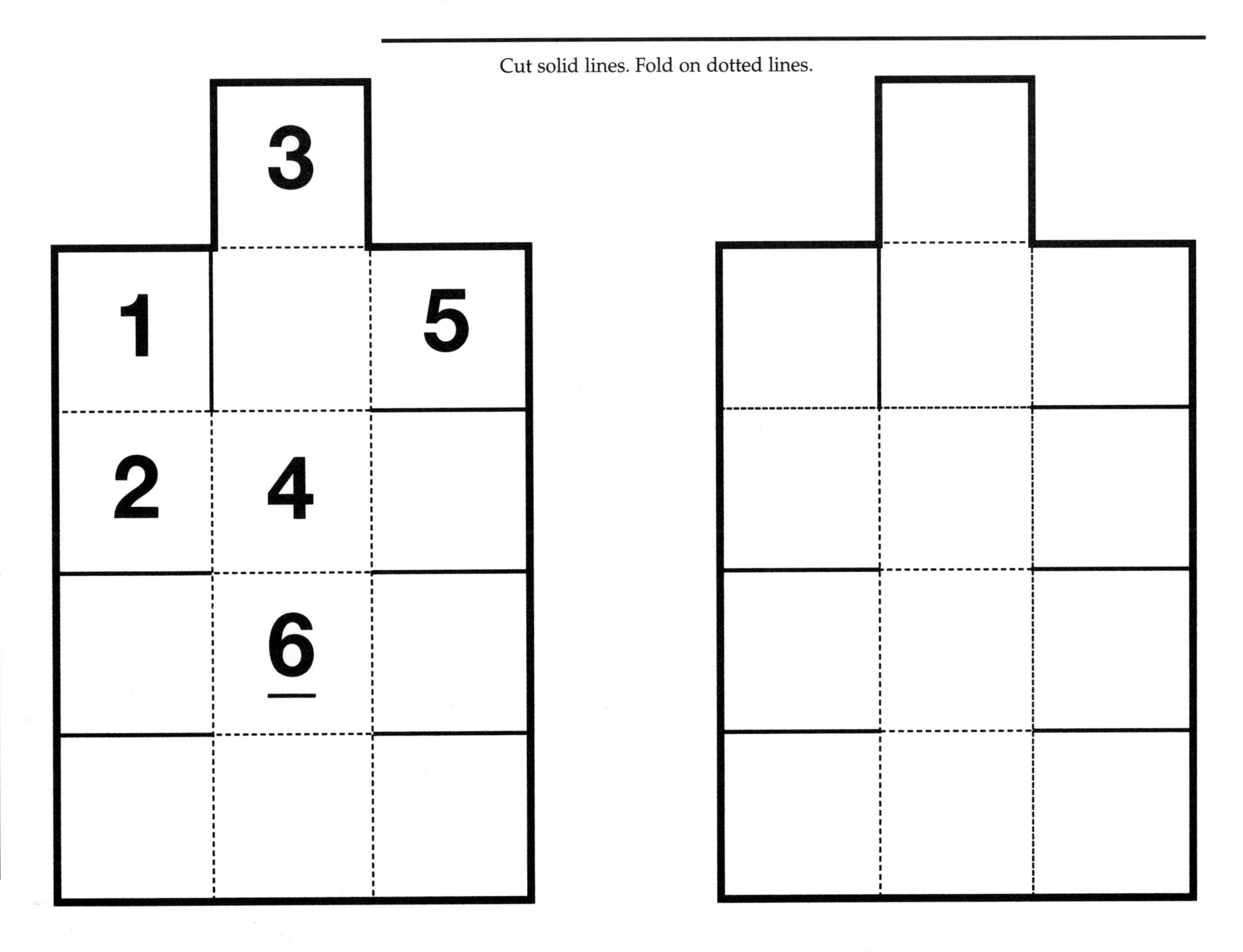

Spinners

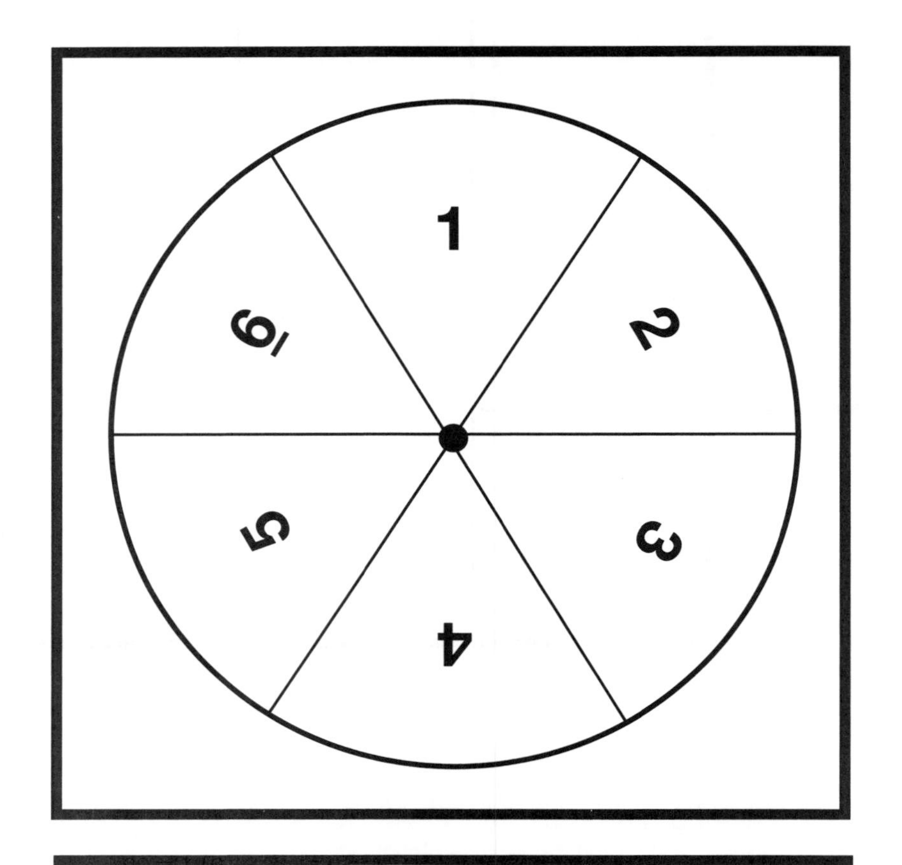

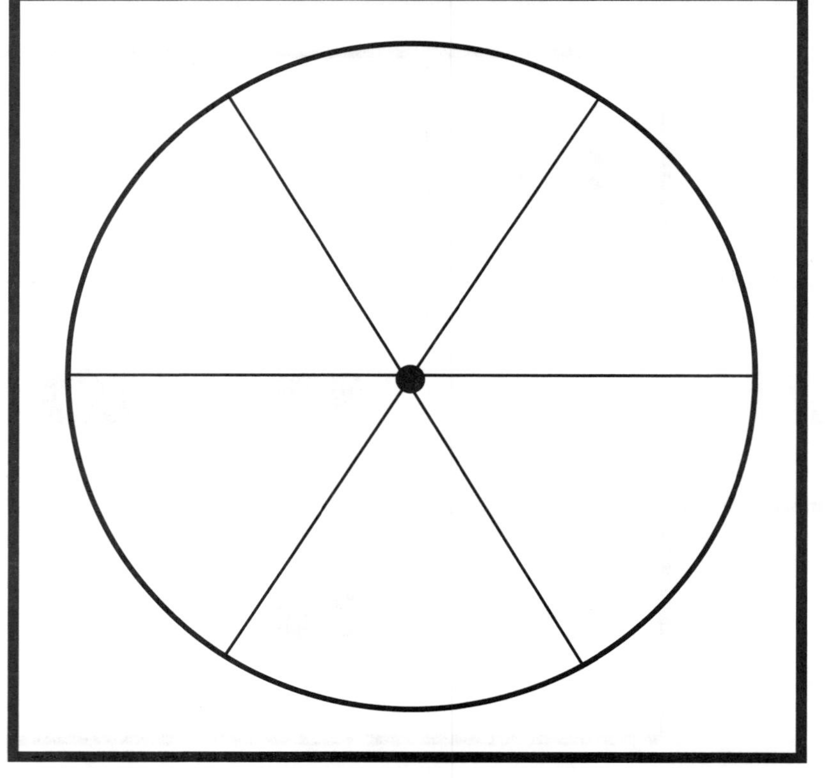

Nimble with Numbers Answer Key

p.16 *Quick Checks 1*

1) 285 2) 327 3) 1480 4) 112 5) 19 6) 9100 7) 276 8) 152r2 9) 3600 10) 700

Go On 731×64

Quick Checks 2

1) 188 2) 564 3) 1380 4) 144 5) 28 6) 6800 7) 325 8) 109r2 9) 8100 10) 3366

Go On $\times, \div; -, \times$

p. 17 *Quick Checks 3*

1) 395 2) 457 3) 1960 4) 104 5) 17 6) 9000 7) 322 8) 142r2 9) 5600 10) 800

Go On $736 \div 2$

Quick Checks 4

1) 326 2) 317 3) 1500 4) 102 5) 26 6) 8400 7) 311 8) 136r2 9) 9600 10) 2871

Go On 3568×2

p. 18 *Quick Checks 5*

1) 296 2) 533 3) 1440 4) 144 5) 18 6) 5600 7) 308 8) 156r2 9) 8400 10) 900

Go On $\div, - ; \div, \times$

Quick Checks 6

1) 185 2) 428 3) 1350 4) 133 5) 27 6) 9600 7) 297 8) 143r1 9) 9000 10) 2673

Go On $567 \div 9$

p. 24 *Choose and Compute*

1) 35×72 2) 56×27 3) 99×56 4) 60×48

5) $2223 \div 39 = 57$ 6) $3276 \div 91 = 36$ 7) $1932 \div 84 = 23$ 8) $1326 \div 17 = 78$

9) $2565 \div 45 = 57$ 10) $3510 \div 45 = 78$ 11) $360 \div 4 \times 45 = 4050$ 12) $82 \times 35 \div 5 \div 7 = 82$

13) $724 \times 13 \div 4 = 2353$ 14) $2296 \div 41 \times 3 = 168$ 15) $2493 \div 9 \times 38 = 10{,}526$

p. 25 *Mental Cross-Number Challenge*

4	6	6	1		8	2	6
2		2	8	5	3		3
2	9		7	1	0		3
1	0	9	6	2		4	9
	3			6	1	7	
5	1	5			5	3	8
	2	0	3		4	2	
2		0		9	0	7	9
7	5	8		3	0		9

Trivia In 1999, French fries had been around for 123 years.

p. 26 *Create and Compute*

Answers will vary. Samples given.

1) 38×5 2) $852 \div 3$ 3) 82×5 4) $325 \div 25$

5) 827×4 6) $8742 \div 6$ 7) 87×42 8) $2784 \div 48$

9) 79×26 10) $9276 \div 12$ 11) 269×7 12) $6792 \div 24$

p. 34 *Number Explorations 1*

1) 40 2) 4 3) 252 4) 4 5) 31, 37, 41, 43 6) Sample: 72 7) 36 or 48 8) $2 \cdot 3^3$

9)

17 16
5
2 12

10)

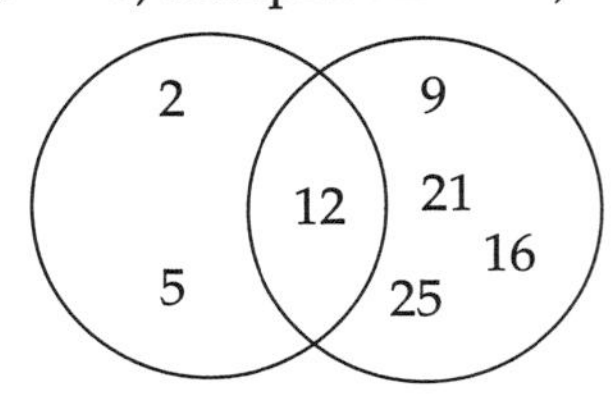

Go On The bottom row is $2^4 \cdot 3^3$.

p. 34 *Number Explorations 2*

1) 60 2) 5 3) 54 4) 9 5) 59, 61, 67 6) Sample: 96 7) 105 or 140 8) $2^3 \cdot 3 \cdot 5$

9)
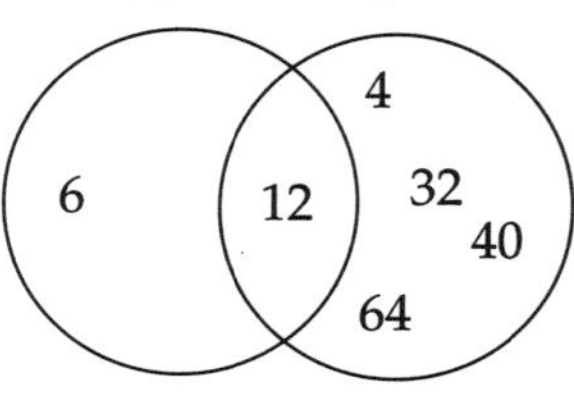

10)
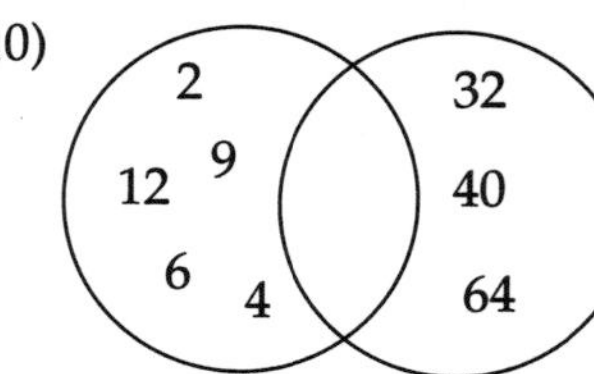

Go On 3 + 5 + 17 = 25; 2 + 5 + 29 = 36

p. 35 *Number Explorations 3*

1) 60 2) 2 3) 72 4) 3 5) 41, 43, 47, 53, 59, 61 6) 40 7) 24 or 48 8) $2^2 \cdot 5^2$

9)
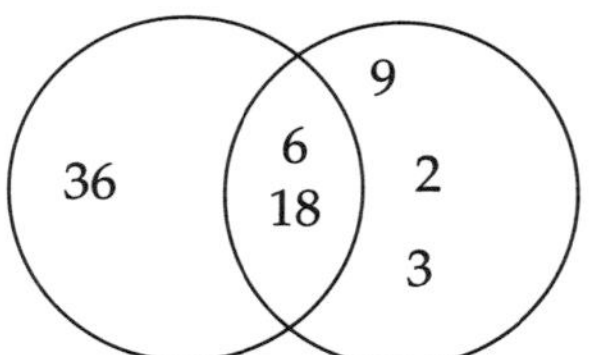

10)
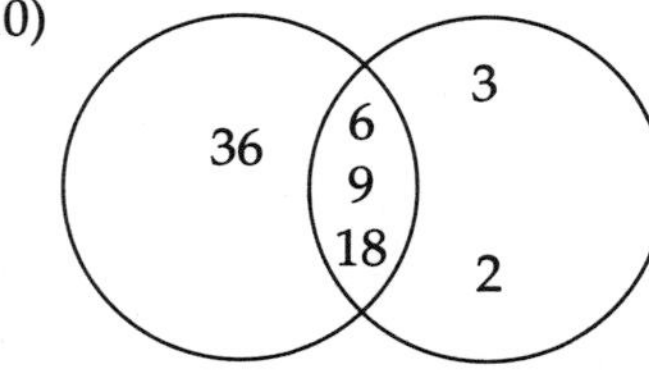

Go On 127, 131, 137, 139, 149, 151, 157, 163, 167, 173

Number Explorations 4

1) 90 2) 2 3) 90 4) 3 5) 83, 89 6) 24 7) 48 8) $2^4 \cdot 5$

9)
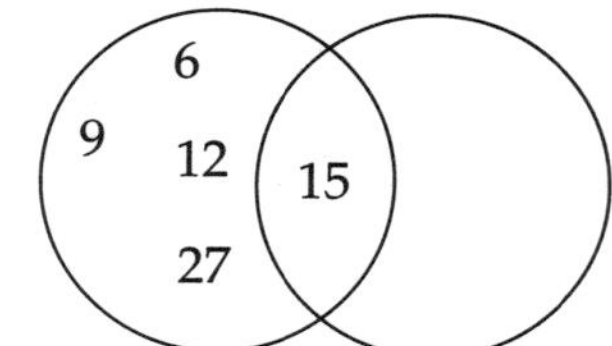

10)
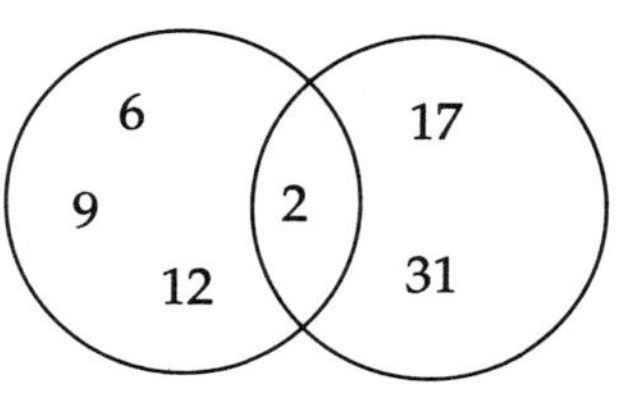

Go On The bottom row is $2^3 \cdot 3^2 \cdot 7$.

p. 36 *Number Explorations 5*

1) 24 2) 4 3) 36 4) 6 5) 71, 73, 79 6) Sample: 84 7) 112 8) 2^6

9)
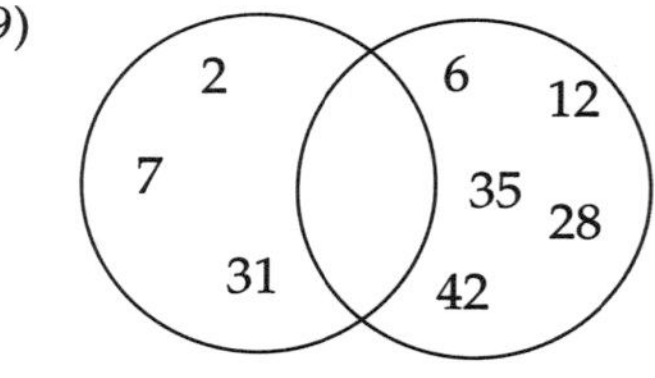

10)
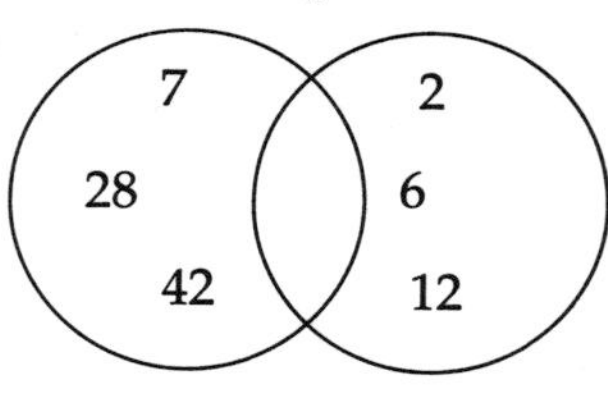

Go On 58, 59

Number Explorations 6

1) 30 2) 2 3z) 60 4) 4 5) 47, 53 6) 12, 24, 36, or 48 7) Sample: 42 8) $2 \cdot 5 \cdot 11$

9)
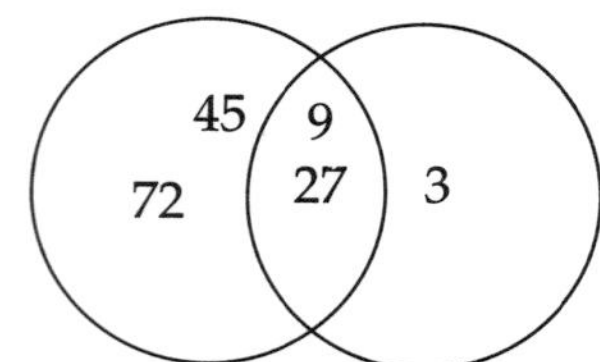

10)

45
40
37
67
3

Go On 179, 181, 191, 193, 197, 199, 211, 223

p. 43 ***Finding Primes***

Answers will vary. Samples are given.

1) Example 2) 3 + 17 3) 7+ 23 4) 5 + 61 5) 5 + 43 6) 3 + 29

7) 3 + 41 8) 3 + 109 9) 3 + 79 10) 5 + 79 11) 5 + 97 12) 3 + 97

The factors resulting from the factor trees are:

13) Example 14) $7 \cdot 2^3$ 15) $2^3 \cdot 3^2$ 16) $2^4 \cdot 5$ 17) $2^2 \cdot 5^2$

18) $2^3 \cdot 3 \cdot 5$ 19) $2^3 \cdot 5^2$ 20) $2^2 \cdot 3 \cdot 13$ 21) $2 \cdot 3^2 \cdot 7$

p. 44 ***Common Factors Practice***

Answers will vary. Sample solution:

3	6	12
21	28	24
7	14	18

p. 45 ***Common Factors Practice Challenge***

Answers will vary. Sample solution:

3	24	15	5
9	36	20	10
6	12	48	4
30	8	16	32

p. 46 ***Using LCM and GCF to Check Multiplication***

1) Example
2) $20 \times 24 = 120 \times 4 = 480$
3) $12 \times 15 = 60 \times 3 = 180$
4) $20 \times 32 = 160 \times 4 = 640$
5) $36 \times 28 = 252 \times 4 = 1008$
6) $28 \times 40 = 280 \times 4 = 1120$
7) $36 \times 45 = 180 \times 9 = 1620$
8) $15 \times 48 = 240 \times 3 = 720$
9) $45 \times 27 = 135 \times 9 = 1215$
10) $14 \times 21 = 42 \times 7 = 294$
11) $36 \times 15 = 180 \times 3 = 540$
12) $25 \times 40 = 200 \times 5 = 1000$
13) $50 \times 30 = 150 \times 10 = 1500$
14) $15 \times 50 = 150 \times 5 = 750$
15) $28 \times 8 = 56 \times 4 = 224$

p. 53 ***Partial Possibilities 1***

1) $\frac{11}{12}$ 2) $1\frac{7}{20}$ 3) $\frac{1}{4}$ 4) $2\frac{5}{7}$ 5) $1\frac{5}{9}$ 6) $\frac{7}{24}$ 7) $1\frac{7}{10}$ 8) $\frac{3}{8}$ 9) $1\frac{1}{5}$ 10) $1\frac{4}{5}$

Go On 3, $2\frac{7}{9}$, $2\frac{5}{9}$, the numbers decrease by $\frac{2}{9}$

Partial Possibilities 2

1) $\frac{5}{6}$ 2) $1\frac{11}{24}$ 3) $\frac{5}{8}$ 4) $3\frac{4}{9}$ 5) $2\frac{1}{6}$ 6) $\frac{7}{12}$ 7) $1\frac{13}{15}$ 8) $\frac{1}{4}$ 9) $1\frac{1}{8}$ 10) $1\frac{5}{9}$

Go On Answers will vary.

p. 54 ***Partial Possibilities 3***

1) $\frac{7}{8}$ 2) $1\frac{8}{21}$ 3) $\frac{1}{6}$ 4) $2\frac{2}{5}$ 5) $2\frac{7}{10}$ 6) $\frac{2}{21}$ 7) $1\frac{7}{12}$ 8) $\frac{5}{12}$ 9) $1\frac{1}{10}$ 10) $1\frac{1}{14}$

Go On $2\frac{1}{12}$, $1\frac{11}{12}$, $1\frac{3}{4}$, the numbers decrease by $\frac{1}{6}$

Partial Possibilities 4

1) $\frac{2}{3}$ 2) $1\frac{7}{15}$ 3) $\frac{1}{3}$ 4) $3\frac{2}{9}$ 5) $1\frac{7}{8}$ 6) $\frac{7}{15}$ 7) $1\frac{7}{12}$ 8) $\frac{3}{14}$ 9) $1\frac{1}{4}$ 10) $1\frac{17}{18}$

Go On Answers will vary.

p. 55 ***Partial Possibilities 5***

1) $\frac{17}{20}$ 2) $1\frac{7}{24}$ 3) $\frac{5}{12}$ 4) $2\frac{2}{7}$ 5) $2\frac{1}{2}$ 6) $\frac{7}{15}$ 7) $1\frac{7}{8}$ 8) $\frac{15}{32}$ 9) $1\frac{5}{7}$ 10) $3\frac{1}{16}$

Go On $1\frac{5}{8}$, $1\frac{5}{24}$, $\frac{19}{24}$, the numbers decrease by $\frac{5}{12}$

Partial Possibilities 6

1) $\frac{19}{24}$ 2) $1\frac{7}{12}$ 3) $\frac{1}{4}$ 4) $3\frac{1}{8}$ 5) $1\frac{9}{10}$ 6) $\frac{9}{35}$ 7) $1\frac{14}{15}$ 8) $\frac{8}{27}$ 9) $1\frac{5}{12}$ 10) $\frac{8}{15}$

Go On Answers will vary.

p. 61 ***Fitting Fractions Practice I***

1) $\frac{6}{1}+\frac{3}{2}=7\frac{1}{2}$ 2) $\frac{2}{6}+\frac{1}{3}=\frac{2}{3}$ 3) $\frac{2}{4}-\frac{1}{3}=\frac{1}{6}$ 4) $\frac{6}{1}-\frac{2}{3}=5\frac{1}{3}$

5) $\frac{2}{3}-\frac{1}{6}=\frac{1}{2}$ 6) $\frac{2}{4}+\frac{1}{3}=\frac{5}{6}$ 7) Sample: $\frac{4}{1}+\frac{2}{3}=4\frac{2}{3}$

p. 62 ***Fitting Fractions Practice II***

1) $\frac{6}{1}\times\frac{3}{2}=9$ 2) $\frac{1}{6}\times\frac{2}{3}=\frac{1}{9}$ 3) $\frac{1}{4}\div\frac{3}{2}=\frac{1}{6}$ 4) $\frac{6}{1}\div\frac{2}{3}=9$

5) Sample: $\frac{1}{2}\times\frac{3}{4}=\frac{3}{8}$ 6) $\frac{4}{1}\times\frac{3}{2}=6$ 7) Sample: $\frac{5}{2}\times\frac{4}{3}=3\frac{1}{3}$ 8) $\frac{2}{6}\div\frac{3}{5}=\frac{5}{9}$

p. 63 ***Creating and Computing Fractions I***

Answers will vary. Samples are given.

1) $\frac{1}{2}+\frac{3}{4}>\frac{8}{12}$ 2) $\frac{1}{4}+\frac{2}{8}<\frac{12}{16}$ 3) $\frac{4}{8}-\frac{1}{2}<\frac{3}{12}$ 4) $\frac{12}{16}-\frac{1}{4}=\frac{1}{2}$

5) $\frac{5}{6}+\frac{1}{3}>\frac{2}{15}$ 6) $\frac{6}{15}+\frac{1}{5}=\frac{3}{5}$ 7) $\frac{2}{3}-\frac{6}{15}<\frac{5}{10}$ 8) $\frac{3}{5}-\frac{2}{10}>\frac{1}{15}$

9) $\frac{3}{6}+\frac{8}{9}>\frac{1}{2}$ 10) $\frac{1}{12}+\frac{2}{6}<\frac{8}{9}$ 11) $\frac{6}{8}-\frac{2}{3}<\frac{9}{12}$ 12) $\frac{6}{8}-\frac{1}{12}=\frac{2}{3}$

p. 64 ***Creating and Computing Fractions II***

Answers will vary. Samples are given.

1) $\frac{1}{2}\times\frac{3}{4}<\frac{5}{8}$ 2) $\frac{4}{5}\times\frac{7}{8}>\frac{1}{2}$ 3) $\frac{2}{5}\div\frac{4}{7}>\frac{3}{8}$ 4) $\frac{1}{8}\div\frac{2}{7}<\frac{3}{4}$

5) $\frac{5}{6}\times\frac{3}{8}<\frac{7}{9}$ 6) $\frac{3}{8}\times\frac{5}{7}>\frac{1}{9}$ 7) $\frac{3}{5}\div\frac{6}{7}>\frac{1}{8}$ 8) $\frac{1}{9}\div\frac{3}{8}>\frac{1}{2}$

9) $\frac{1}{4}\times\frac{2}{9}<\frac{5}{6}$ 10) $\frac{2}{5}\times\frac{9}{10}=\frac{9}{25}$ 11) $\frac{4}{6}\div\frac{2}{5}>\frac{1}{9}$ 12) $\frac{1}{6}\div\frac{2}{9}<\frac{4}{5}$

p. 72 ***Partial Possibilities 7***

1) 4.023 2) Sample: 0.285 3) 2.45 × 1.6 = 3.92 4) $\frac{1}{4}$, 0.37, $\frac{3}{5}$, 0.62

5) 2.25 6) 53.9 7) 499.97 8) $<$

9) 40.608 10) 43.2

Go On 20.83, 25.21, 29.59; the numbers increase by 4.38.

Partial Possibilities 8

1) 16.037 2) Sample: 0.5123 3) 37.1 × 8.2 = 304.22 4) $\frac{1}{3}$, 0.4, $\frac{4}{5}$, 0.85

5) 4 6) 728 7) 190.66 8) $>$

9) 28.348 10) 75.8

Go On 0.258 × 0.47

p. 73 ***Partial Possibilities 9***

1) 18.009 2) Sample: 0.437 3) 6.32 × 4.8 = 30.336 4) 0.17, $\frac{1}{5}$, $\frac{2}{4}$, 0.67

5) 2.5 6) 0.459 7) 332.48 8) $>$

9) 41.965 10) 56.4

Go On 17.1, 15.7, 14.3; the numbers decrease by 1.4.

Partial Possibilities 10

1) 3.081 2) Sample: 0.8372 3) 4.8 × 0.975 = 4.68 4) 0.51, 0.65, $\frac{2}{3}$, $\frac{3}{4}$

5) 4.7 6) 61.2 7) 318.97 8) $>$

9) 27.434 10) 49.72

Go On 763 ÷ 0.1

p. 74 ***Partial Possibilities 11***

1) 9.052 2) Sample: 0.642 3) $7.4 \times 3.65 = 27.01$ 4) 0.73, 0.80, $\frac{5}{6}$, $\frac{7}{8}$

5) 3.6 6) 1876 7) 402.07 8) >

9) 29.348 10) 7.2

Go On 19.12, 38.24, 76.48; multiply by 2.

Partial Possibilities 12

1) 17.016 2) Sample: 0.2663 3) $96.1 \times 5.3 = 509.33$ 4) $\frac{1}{3}$, 0.37, $\frac{2}{5}$, 0.52

5) 7.5 6) 3.48 7) 163.24 8) <

9) 3.6394 10) 314.5

Go On 0.5, 2, 2.5; 1.3, 2.2, 2.5

p. 80 ***Decimal Trails I***

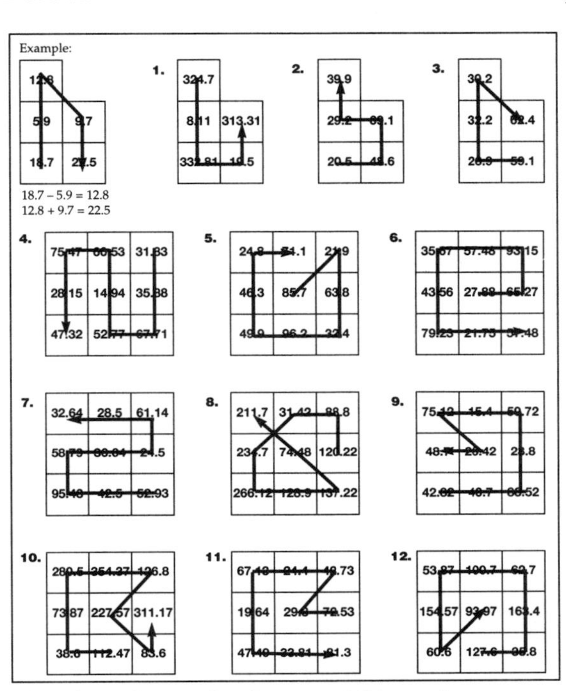

Note: Since these paths alternate addition and subtraction, arrow can be going in opposite direction.

p. 81 ***Decimal Trails II, Trails may vary***

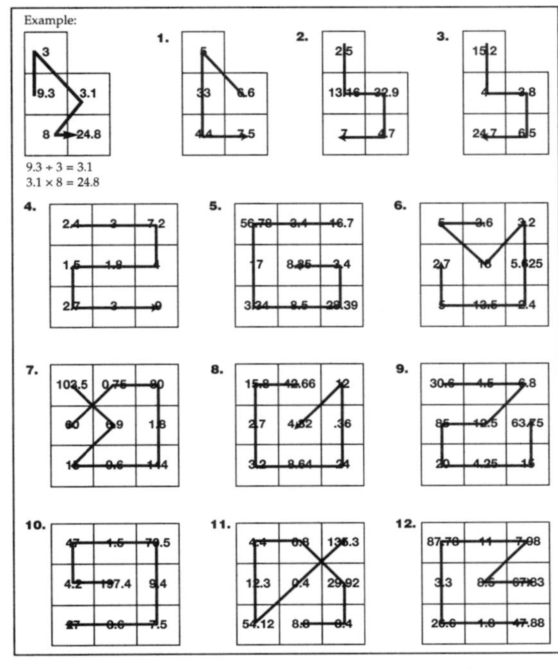

Note: Paths may vary somewhat since order of factors or divisors can be reversed.

p. 82 ***Fitting Decimals I***

Terms may vary although the final answer should be as shown.

1) 75.32 + 6.4 = 81.72 78.65 – 2.4 = 76.25
   6.42 + 75.5 = 81.92 76.5 – 9.24 = 67.26

2) 6.59 + 34.8 = 41.39 34.6 – 9.78 = 24.82
   0.489 + 32.6 = 33.089 93.1 – 88.64 = 4.46

3) 23.98 + 5.7 = 29.68 23.5 – 6.87 = 16.63
   5.18 + 234.7 = 239.88 3.957 – 2.8 = 1.157

4) 0.751 + 9.63 = 10.381 9.7 – 0.136 = 9.564
   7.41 + 96.83 = 104.24 92.7 – 13.06 = 79.64

5) 2.48 + 0.769 = 3.249 24.6 – 9.8 = 14.8
   46.8 + 25.39 = 72.19 91.2 – 86.84 = 4.36

p. 83 ***Fitting Decimals II***

1) $84.1 \times 0.65 = 54.67$ $8.65 \div 1.4 = 6.18$
   $83.41 \times 6.5 = 542.17$ $86.5 \div 6.14 = 14.09$

2) $46.79 \times 3.5 = 163.77$ $34.75 \div 0.97 = 35.82$
   $8.479 \times 3.25 = 27.56$ $36.45 \div 9.7 = 3.76$

3) $43.6 \times 2.58 = 112.49$ $235 \div 0.86 = 273.26$
   $9.5 \times 23.68 = 224.96$ $23.75 \div 8.6 = 2.76$

4) $67.5 \times 0.942 = 63.59$ $39.75 \div 0.24 = 165.63$
   $91.942 \times 7.5 = 689.57$ $9.75 \div 0.24 = 40.63$

5) $1.36 \times 0.547 = 0.74$ $183.4 \div 7.6 = 24.13$
   $93.47 \times 1.46 = 136.47$ $11.34 \div 0.76 = 14.92$

p. 84 ***Magic Squares***

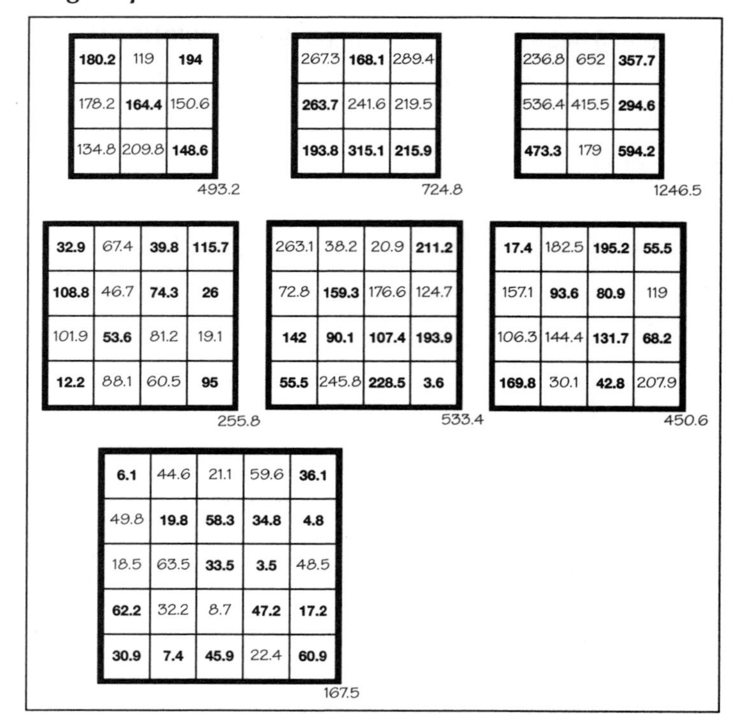

180.2	119	194
178.2	164.4	150.6
134.8	209.8	148.6

493.2

267.3	168.1	289.4
263.7	241.6	219.5
193.8	315.1	215.9

724.8

236.8	652	357.7
536.4	415.5	294.6
473.3	179	594.2

1246.5

32.9	67.4	39.8	115.7
108.8	46.7	74.3	26
101.9	53.6	81.2	19.1
12.2	88.1	60.5	95

255.8

263.1	38.2	20.9	211.2
72.8	159.3	176.6	124.7
142	90.1	107.4	193.9
55.5	245.8	228.5	3.6

533.4

17.4	182.5	195.2	55.5
157.1	93.6	80.9	119
106.3	144.4	131.7	68.2
169.8	30.1	42.8	207.9

450.6

6.1	44.6	21.1	59.6	36.1
49.8	19.8	58.3	34.8	4.8
18.5	63.5	33.5	3.5	48.5
62.2	32.2	8.7	47.2	17.2
30.9	7.4	45.9	22.4	60.9

167.5

p. 92 ***Perky Portions 1***

1) 60% 2) $\frac{13}{20}$ 3) 100 4) 72 5) 2.5% 6) 16.6 7) 180

8) 1.7%, $\frac{1}{6}$, 0.63, 82% 9) $15.59 10) $53 Go On 7.56 Answers will vary.

Perky Portions 2

1) 62.5% 2) $\frac{19}{50}$ 3) 80 4) 7 5) 12.5% 6) 50% 7) 40

8) 37%, $\frac{3}{7}$, 0.73, $\frac{7}{3}$ 9) $39.67 10) $133.75 Go On Answers will vary.

p. 93 ***Perky Portions 3***

1) 80% 2) $\frac{17}{20}$ 3) 24 4) 88 5) 9.2% 6) 56 7) 200

8) 0.37, $\frac{3}{8}$, 0.41, 52% 9) $13.82 10) $84.80 Go On Answers will vary.

Perky Portions 4

1) 37.5% 2) $\frac{7}{20}$ 3) 80 4) 6 5) 36.5% 6) 80% 7) 150

8) 0.56, $\frac{5}{8}$, 65%, $\frac{5}{6}$ 9) $32.23 10) $187.25 Go On 15

p. 94 ***Perky Portions 5***

1) 40% 2) $\frac{9}{20}$ 3) 45 4) 32 5) 8.7% 6) 80% 7) 60

8) 40%, $\frac{9}{17}$, $\frac{2}{3}$, 1.27 9) $19.44 10) $212 Go On Answers will vary.

Perky Portions 6

1) 87.5% 2) $\frac{47}{100}$ 3) 44 4) 100 5) 76.2% 6) 80% 7) 70

8) $\frac{8}{17}$, $\frac{5}{8}$, 64%, 1.03 9) $53.21 10) $262.15 Go On Answers will vary.

p. 102 *Ordered Mazes*

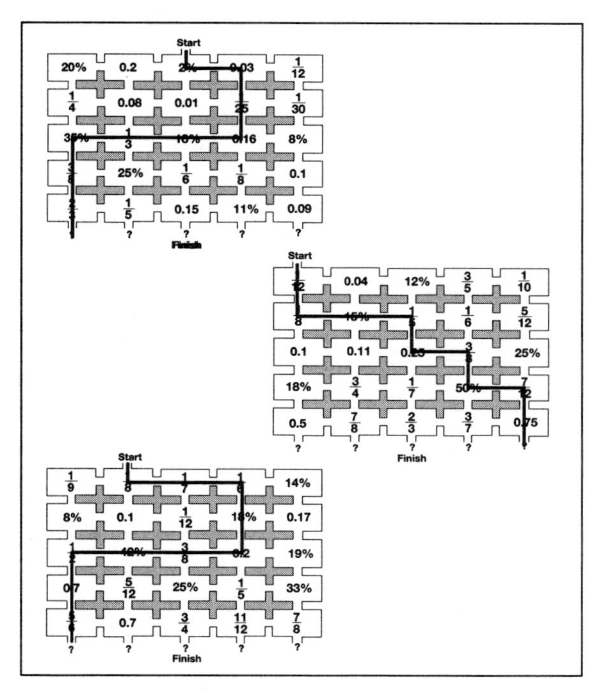

p. 103 *Relating Quantities**

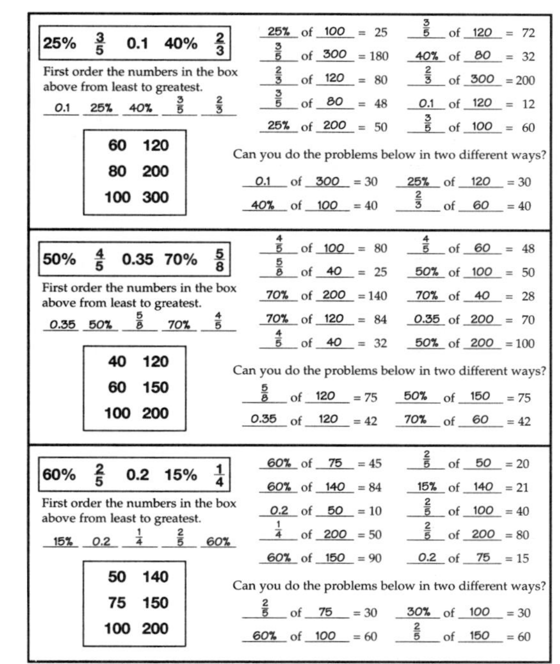

25% $\frac{3}{5}$ 0.1 40% $\frac{2}{3}$

First order the numbers in the box above from least to greatest.

0.1, 25%, 40%, $\frac{3}{5}$, $\frac{2}{3}$

60	120
80	200
100	300

$\frac{3}{5}$ of 120 = 72
40% of 80 = 32
$\frac{2}{3}$ of 300 = 200
0.1 of 120 = 12
$\frac{3}{5}$ of 100 = 60

25% of 100 = 25
$\frac{3}{5}$ of 300 = 180
$\frac{2}{3}$ of 120 = 80
$\frac{3}{5}$ of 80 = 48
25% of 200 = 50

Can you do the problems below in two different ways?

0.1 of 300 = 30 — 25% of 120 = 30
40% of 100 = 40 — $\frac{2}{3}$ of 60 = 40

50% $\frac{4}{5}$ 0.35 70% $\frac{5}{8}$

First order the numbers in the box above from least to greatest.

0.35, 50%, $\frac{5}{8}$, 70%, $\frac{4}{5}$

40	120
60	150
100	200

$\frac{4}{5}$ of 100 = 80
$\frac{5}{8}$ of 40 = 25
70% of 200 = 140
70% of 120 = 84
$\frac{4}{5}$ of 40 = 32

$\frac{4}{5}$ of 60 = 48
50% of 100 = 50
70% of 40 = 28
0.35 of 200 = 70
50% of 200 = 100

Can you do the problems below in two different ways?

$\frac{5}{8}$ of 120 = 75 — 50% of 150 = 75
0.35 of 120 = 42 — 70% of 60 = 42

60% $\frac{2}{5}$ 0.2 15% $\frac{1}{4}$

First order the numbers in the box above from least to greatest.

15%, 0.2, $\frac{1}{4}$, $\frac{2}{5}$, 60%

50	140
75	150
100	200

60% of 75 = 45
60% of 140 = 84
0.2 of 50 = 10
$\frac{1}{4}$ of 200 = 50
60% of 150 = 90

$\frac{2}{5}$ of 50 = 20
15% of 140 = 21
$\frac{2}{5}$ of 100 = 40
$\frac{2}{5}$ of 200 = 80
0.2 of 75 = 15

Can you do the problems below in two different ways?

$\frac{2}{5}$ of 75 = 30 — 30% of 100 = 30
60% of 100 = 60 — $\frac{2}{5}$ of 150 = 60

* There may be other solutions for some of the examples.

p. 104 *Cross Number Puzzle*

2	3	■	■	1	2	5	■	■
■	3	2	2	5	■	5	4	6
5	6	■	■	4	8	■	■	4
2	■	9	6	■	■	7	6	■
8	5	5	■	■	1	■	5	6
■	9	■	1	3	6	■	8	■

Trivia: 95%

p. 112 *Negative Notions 1*

1) 3 2) 10 3) $^-10$ 4) $^-45$ 5) +, – 6) –, – 7) $^-21 + 4 + ^-2$ 8) $^-8 + ^-9$

9) Sample: $\boxed{^-6} \oplus \boxed{^-4} \ominus \boxed{8}$ 10) Sample: $\boxed{^-4} \ominus \boxed{8} \ominus \boxed{^-6}$

Go On Samples: $\boxed{^-6} \oplus \boxed{^-4} \oplus \boxed{3} \oplus \boxed{8}$; $(\boxed{8} \ominus \boxed{6}) \oslash (\boxed{^-4} \ominus \boxed{3})$

Negative Notions 2

1) $^-5$ 2) $^-14$ 3) 15 4) $^-9$ 5) +, – 6) –, + 7) $8 + ^-14 + ^-6 + 5$ 8) $^-5 + ^-6 + ^-7$

9) Sample: $\boxed{^-3} \ominus \boxed{4} \ominus \boxed{7}$ 10) Sample: $\boxed{2} \oplus \boxed{4} \ominus \boxed{7}$

Go On Samples: $\boxed{4} \ominus \boxed{^-3} \ominus \boxed{2} \ominus \boxed{7}$; $\boxed{7} \ominus \boxed{^-3} \ominus \boxed{2} \ominus \boxed{^-4}$

p. 113 *Negative Notions 3*

1) $^-15$ 2) 4 3) $^-27$ 4) 42 5) –, – 6) +, – 7) $^-7 + 5 + ^-8$ 8) $^-11 + ^-13$

9) Sample: $\boxed{^-7} \otimes \boxed{^-5} \ominus \boxed{1}$ 10) Sample: $\boxed{1} \ominus \boxed{4} \ominus \boxed{^-5}$

Go On Samples: $\boxed{4} \ominus \boxed{^-7} \oplus \boxed{1} \oplus \boxed{^-5}$; $\boxed{^-5} \ominus \boxed{4} \ominus \boxed{^-7} \ominus \boxed{1}$

Negative Notions 4

1) 0 2) 14 3) 1 4) $^-5$ 5) +, –, 6) –, – 7) $^-9 + 15 + ^-7 + 4$ 8) $^-8 + ^-10 + ^-12$

9) Sample: $\boxed{^-6} \otimes \boxed{5} \oslash \boxed{2}$ 10) Sample: $\boxed{2} \ominus \boxed{^-6} \ominus \boxed{5}$

Go On Samples: $\boxed{2} \otimes \boxed{5} \ominus \boxed{^-9} \oplus \boxed{^-6}$; $\boxed{^-6} \ominus \boxed{5} \ominus \boxed{2} \otimes \boxed{^-9}$

p. 114 *Negative Notions 5*

1) -4 2) -5 3) 1 4) 48 5) +, – 6) –, – 7) 4 + -3 + 6 8) -6 + -8

9) Sample: [-8] ⊖ [10] ⊖ [-5] 10) Sample: [3] ⊗ [10] ⊕ [-5]

Go On Samples: [-8] ⊖ [-5] ⊕ [3] ⊕ [10] ; [-8] ⊗ [-5] ⊕ [10] ⊕ [3]

Negative Notions 6

1) -8 2) -14 3) 18 4) 9 5) –, – 6) +, – 7) -7 + 13 + -8 + 2 8) -3 + -5 + -7

9) Sample: [9] ⊕ [-4] ⊖ [-7] 10) Sample: [6] ⊖ [9] ⊖ [-7]

Go On Samples: [-7] ⊖ [9] ⊖ [-4] ⊗ [6] ; [6] ⊖ [9] ⊖ [-4] ⊕ [-7]

p. 121 *Finding Sums and Differences*

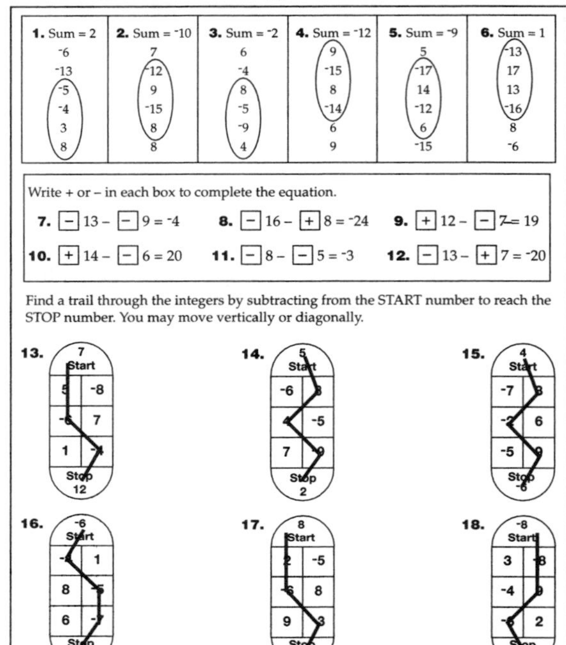

1. Sum = 2	2. Sum = -10	3. Sum = -2	4. Sum = -12	5. Sum = -9	6. Sum = 1
-6	7	6	9	5	-13
-13	-12	-4	-15	-17	17
-5	9	8	8	14	13
-4	-15	-5	-14	-12	-16
3	8	-9	6	6	8
8	8	4	9	-15	-6

Write + or – in each box to complete the equation.

7. [–] 13 – [–] 9 = -4 8. [–] 16 – [+] 8 = -24 9. [+] 12 – [–] 7 = 19

10. [+] 14 – [–] 6 = 20 11. [–] 8 – [–] 5 = -3 12. [–] 13 – [+] 7 = -20

Find a trail through the integers by subtracting from the START number to reach the STOP number. You may move vertically or diagonally.

13. 7 Start; 5, -8; -6, 7; 1, -4; Stop 12

14. 5 Start; -6, 8; 4, -5; 7, -9; Stop 2

15. 4 Start; -7, 8; -2, 6; -5, -9; Stop -6

16. -6 Start; -4, 1; 8, -5; 6, -7; Stop 10

17. 8 Start; 2, -5; -6, 8; 9, -3; Stop 15

18. -8 Start; 3, -8; -4, 9; -6, 2; Stop -3

p. 122 *Finding Tic-Tac-Toes I*

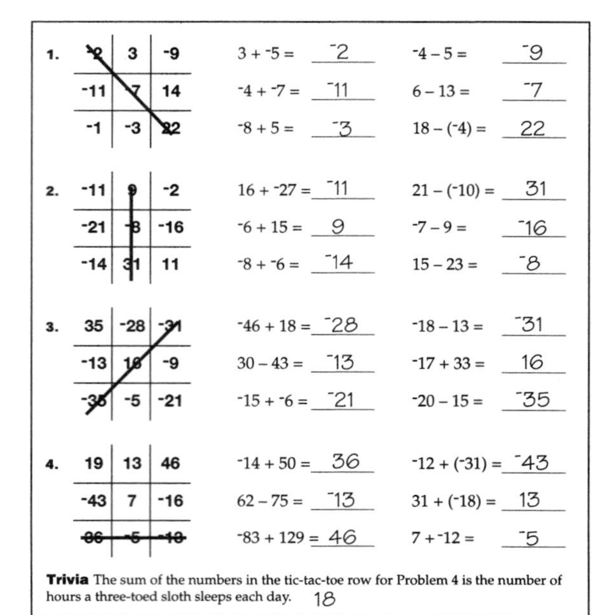

1.

-2	3	-9
-11	-7	14
-1	-3	22

3 + -5 = -2 -4 – 5 = -9
-4 + -7 = -11 6 – 13 = -7
-8 + 5 = -3 18 – (-4) = 22

2.

-11	9	-2
-21	-8	-16
-14	31	11

16 + -27 = -11 21 – (-10) = 31
-6 + 15 = 9 -7 – 9 = -16
-8 + -6 = -14 15 – 23 = -8

3.

35	-28	-31
-13	16	-9
-35	-5	-21

-46 + 18 = -28 -18 – 13 = -31
30 – 43 = -13 -17 + 33 = 16
-15 + -6 = -21 -20 – 15 = -35

4.

19	13	46
-43	7	-16
-86	-5	-10

-14 + 50 = 36 -12 + (-31) = -43
62 – 75 = -13 31 + (-18) = 13
-83 + 129 = 46 7 + -12 = -5

Trivia The sum of the numbers in the tic-tac-toe row for Problem 4 is the number of hours a three-toed sloth sleeps each day. 18

p. 123 *Finding Tic-Tac-Toes II*

1.

-60	-42	-27
-18	27	-48
66	18	60

6 • -8 = -48 (5 • -3) • -4 = 60
-3 • -9 = 27 -2 • (11 – -3) = 66
-7 • 6 = -42 (36 ÷ 4) • 2 = -18

2.

-42	36	-4
60	-4	24
-28	-36	28

-5 • -12 = 60 (63 ÷ -9) • -4 = 28
9 • -4 = -36 -48 ÷ (-2 • -6) = -4
-28 ÷ -7 = 4 (42 ÷ -7) • 7 = -42

3.

-56	12	-25
63	-32	4
-63	-12	-64

7 • -8 = -56 60 ÷ (-30 ÷ -2) = 4
-50 ÷ 2 = -25 (8 • -6) ÷ 4 = -12
(36 ÷ -4) • -7 = 63 -4 – (48 ÷ 3) = -64

4.

-72	24	30
-30	-3	-20
-12	108	72

-45 ÷ 15 = -3 80 ÷ (-36 ÷ 9) = -20
(-8 • -3) • -3 = -72 (8 • -6) ÷ 4 = -12
9 • (-36 ÷ -3) = 108 (-15 • 4) ÷ -2 = 30

Trivia The sum of the numbers in the tic-tac-toe row for Problem 4 is the number of inches across the eye of a giant squid. 15

p. 124 *Integer Choices*

Answers will vary.

p. 133 *Equation Station 1*

1) 32 2) 11 3) $x = 5$ 4) $x = 7$ 5) $x = 36$
6) $x = 0.3$ 7) $x = 4$ 8) $x = 21$ 9) $x = 5$ 10) $x = 4$
Go On

x	y	$x + y$	$x \times y$
20	32	52	640
22	18	40	396

Equation Station 2

1) 44 2) 43 3) $x = 15$ 4) $x = 14$ 5) $x = 72$
6) $x = 2.4$ 7) $x = 21$ 8) $x = 8$ 9) $x = {}^-3$ 10) $x = 3$
Go On Answers will vary.

p. 134 *Equation Station 3*

1) 59 2) 26 3) $x = 31$ 4) $x = 9$ 5) $x = 45$
6) $x = 170$ 7) $x = 9$ 8) $x = 10$ 9) $x = 10$ 10) $x = 3$
Go On

B	C	$B - C$	$B \div C$
512	32	480	16
80	20	60	4

Equation Station 4

1) 137 2) 122 3) $x = 71$ 4) $x = 12$ 5) $x = 63$
6) $x = 15$ 7) $x = 15$ 8) $x = 6$ 9) $x = 4$ 10) $x = 5$
Go On Answers will vary.

p. 135 *Equation Station 5*

1) 58 2) ${}^-5$ 3) $x = {}^-15$ 4) $x = 8$ 5) $x = 75$
6) $x = 0.3$ 7) $x = 12$ 8) $x = 12$ 9) $x = 4$ 10) $x = 2$
Go On

M	N	$M \times N$	$M - N$
55	37	2035	18
50	12	600	38

Equation Station 6

1) 14 2) 122 3) $x = 52$ 4) $x = 6$ 5) $x = 88$
6) $x = 7$ 7) $x = 8$ 8) $x = 2$ 9) $x = {}^-2$ 10) $x = 2$
Go On Answers will vary.

p. 143 *Symbolizing Number Riddles*

1) $x + 17 = 25; x = 8$ 2) $x - 9 = 15; x = 24$ 3) $x - 8 = 30; x = 38$
4) $x + 10 = 51; x = 41$ 5) $3x = 39; x = 13$ 6) $\frac{x}{4} = 20; x = 80$
7) $\frac{x}{2} = 36; x = 72$ 8) $6x = 54; x = 9$ 9) $2x + 6 = 20; x = 7$
10) $5x - 4 = 46; x = 10$ 11) $8x - 15 = 73; x = 11$ 12) $3x + 9 = 60; x = 17$
13) $10 + \frac{x}{3} = 21; x = 33$ 14) $28 + \frac{x}{4} = 40; x = 48$ 15) $\frac{x}{5} - 7 = 9; x = 80$
16) $\frac{x}{2} - 17 = 13; x = 60$ 17) $\frac{x + 9}{5} = 3; x = 6$ 18) $\frac{x - 15}{3} = 4; x = 27$

p. 144 *Geographic Discoveries via Algebra*

1) $x = 3$ 2) $x = 2$ 3) $x = 7$ 4) $x = 2$ 5) $x = 4$
6) $x = 2$ 7) $x = 6$ 8) $x = 7$ 9) $x = 4$ 10) $x = 8$
11) $x = 7$ 12) $x = 5$ 13) $x = 9$ 14) $x = 6$ 15) $x = 2$
16) Canada 17) Indonesia